T0332549

THE GLOBAL PALEOMAGNETIC DATABASE

THE GLOBAL PALEOMAGNETIC DATABASE

Design, Installation and Use with ORACLE

Edited by
JO LOCK and MICHAEL W. McELHINNY
Gondwana Consultants Pty Ltd, Fishing Point NSW, Australia

Reprinted from Surveys in Geophysics, Volume 12, Nos. 4–5 (1991)

KLUWER ACADEMIC PUBLISHERS
DORDRECHT / BOSTON / LONDON

ISBN 0-7923-1327-5

Published by Kluwer Academic Publishers,
P.O. Box 17, 3300 AA Dordrecht, The Netherlands.

Kluwer Academic Publishers incorporates
the publishing programmes of
D. Reidel, Martinus Nijhoff, Dr W. Junk and MTP Press.

Sold and distributed in the U.S.A. and Canada
by Kluwer Academic Publishers,
101 Philip Drive, Norwell, MA 02061, U.S.A.

In all other countries, sold and distributed
by Kluwer Academic Publishers Group,
P.O. Box 322, 3300 AH Dordrecht, The Netherlands.

Printed on acid-free paper

All Rights Reserved
© 1991 Kluwer Academic Publishers
No part of the material protected by this copyright notice may be reproduced or
utilized in any form or by any means, electronic or mechanical,
including photocopying, recording or by any information storage and
retrieval system, without written permission from the copyright owner.

Printed in the Netherlands

SUMMARY

PREFACE

At the IUGG General Assembly in Vancouver, the IAGA Division I Working Groups in Paleomagnetism and Rock Magnetism discussed the need for the development of global databases covering all aspects of data used by researchers in the two working groups. At its plenary session at the close of the Assembly IAGA passed a formal resolution as follows sponsored by Division I:

"IAGA, *recognising* the importance of paleomagnetic data for studies of the tectonic history of the earth and for understanding the geomagnetic field and *noting* the lack of any international coordination in preserving and compiling such data, urges support for the compilation of both regional and global databases and for international workshops to coordinate the merging of such databases in order to facilitate the use of paleomagnetic data by research workers."

Following that meeting it was agreed to try to set up the first of these databases for Directions and Pole Positions through funding from various agencies in different countries. The project has thus been supported by the following countries and agencies.

United States   – National Science Foundation (Grant to R. Van der Voo, University of Michigan)
United Kingdom – Natural Environment Research Council (Direct grant to M. W. McElhinny)
Australia   – Bureau of Mineral Resources (Contract to Gondwana Consultants)
Germany   – Deutsche Forschungsgemeinschaft (Grant to H. C. Soffel)
France   – Institut National des Science de l'Univers (Grant to V. Courtillot on behalf of all French laboratories)
Holland   – Dutch Science Foundation (Grant to J. D. A. Zijderveld)
Switzerland   – Swiss Nationalfonds (Grant to W. Lowrie)
Japan   – Rock and Paleomagnetic Research Group (M. Kono)
Soviet Union   – VNIGRI Institute, Leningrad (A. N. Khramov)

Although global catalogues have been developed over many years by Irving and by McElhinny (Geophysical Journal Pole Lists and the Ottawa Series from the Earth Physics Branch), the need was for the development of a modern relational database using the most up-to-date system. It was crucial to have a system that allowed sophisticated queries through the full range of relational operators and that also allowed for future expansion. After full consideration of all the facts we came to the conclusion that the ORACLE Relational Database Management System was the most appropriate PC system. This view was endorsed at the first

International Symposium in Czechoslovakia 27 June–2 July, 1988 to consider the future development of the IAGA Division I databases in Paleomagnetism and Rock Magnetism. Reports of this meeting and introductory aspects of the Global Database have been published in papers by Van der Voo and McElhinny (1989) and McElhinny and Lock (1990a,b).

Our purpose in producing this document is not just to explain the structure and content of the Global Paleomagnetic Database (GPMDB) but to provide a working outline of the ORACLE Relational Database Management System (RDBMS) for use with the Database on a Personal Computer (PC) under MS-DOS. The GPMDB can also be installed using ORACLE on a mainframe computer and all the programs we have developed for use with the various ORACLE utilities should operate just as well on a mainframe. When you purchase ORACLE the software comes with a set of 20 volumes that is quite overwhelming for the uninitiated, as we were when we first started. Not only does one have to obtain an understanding of SQL, now designated the international language for relational databases, but one has to cope with becoming the manager of an extremely powerful software system. We explain in the first section why ORACLE was chosen. After two years of working with ORACLE, we are quite convinced that the paleomagnetic community made the correct choice.

The ORACLE documentation is largely based around business applications but the system is very applicable to scientific databases. In this document we have provided a summary of how to use ORACLE with the GPMDB. All the examples we give are based around that database and we have, to a large extent, filtered the 20 volumes of ORACLE down to this one volume. Much of our experience with developing the GPMDB, including some of the pitfalls of SQL and ORACLE, are included here to help others avoid some of the traumas we went through.

This document assumes that many paleomagnetists will be inexperienced PC users. To assist those who will be using a PC with MS-DOS for the first time we have included those essential aspects of MS-DOS that are necesary both to install and maintain the ORACLE RDBMS software. The tutorials are designed to give new ORACLE users a good basic knowledge of SQL as applied to the GPMDB. As new users become more experienced they will find the ORACLE documentation easier to comprehend. Advanced users will be able to find what they reguire in the ORACLE documenation. You will undoubtedly have a lot of fun and the world's paleomagnetic data will be at your fingertips! In many examples an author name is required. To ensure that no one is offended we have used 'McElhinny' as that author name.

The document is divided into 5 sections not all of which necessarily have to be read to get started. Section I is about the design of the database and should be read by everyone. Section II describes the PC system requirements and the ORACLE software that is used by the database. Once you have purchased these items, obviously there will be little need to refer back to this section again. At each installation one person will need to be the designated Database Administrator (DBA). If you are on your own you will be that person. Section III is addressed to the DBA in your group and describes the installation and maintenance of the database. Once loaded, users will mostly refer to Section IV and V on using the database. Section IV will be the easiest place to start by using the Menu Application we have developed. Those who would like to become more expert users of SQL*Plus and to develop their own queries and programs or even their own prlvate databases wlll need to refer to Section V in some detail.

We have prepared a special set of pre-packaged programs for use with the ORACLE Database System. These programs and the database are available on $5\frac{1}{4}$ inch floppy diskettes from the World Data Centers. To obtain copies write to:

> World Data Center A,
> National Geophysical Data Center, NOAA
> 325 Broadway, E/GC
> BOULDER, Colorado 80303–3328
> U.S.A.

At the time of going to press WDC-A charges a copying fee of $30 per diskette plus a handling charge of $10 plus another $10 for non U.S. requests.

The Editors

DOCUMENT SYNTAX

There are several conventions we have used with respect to the syntax used in this document as follows:

Messages (including error messages) that might be given by ORACLE or by MS-DOS are given in quotes "........".

Function keys on the computer keyboard are given in square brackets, such as [F2] or [Ctrl][Alt][Del].

Options related to SQL*Plus commands are given in square brackets. For example Command[.option] indicates that the Command may be issued alone or with an option as shown.

All SQL and SQL*Plus commands have been given in capitals. Table names, column names and view names are given in capitals.

Trademarks
The following trademarks appear in the text and are acknowledged here:

dBASE, dBASE III, dBASE IV are trademarks of Ashton-Tate.
IBM, IBM PC, PC AT, SQL/DS and DB2 are trademarks of International Business Machines Inc.
MS-DOS is a trademark of Microsoft Corporation.
Norton Utilities and Speed Disk are trademarks of Peter Norton Computing Inc.
ORACLE, SQL*Plus, SQL*Forms, SQL*Menu, SQL*Loader are trademarks of the Oracle Corporation.
Wordstar is a trademark of MicroPro International Corporation.
XTreePro and XtreeGold are trademarks of the Xtree Company.

I. DATABASE DESIGN

1. Relational Databases

It is important to distinguish between the terms *data catalogue* and *database*. In modern terms *database* now refers to data stored in a manner that optimises retrieval of the data with a database management system. At present the most efficient form of such systems for the type of data developed from paleomagnetic studies is the *Relational Database Management System (RDBMS)*.

The development of databases commences with an understanding of the data items to be stored and the associations between those data items. Some simple examples from paleomagnetism illutrate the different types of associations between data items.

RESULT # < ------------------- > POLE POSITION

1:1 Association

Each result number has only one pole position, and each pole position only refers to one result number. There is a one-to-one association between them.

ROCKUNIT # < ----------------≫ PALEOMAGNETIC RESULT

1:N Association

Each rockunit number may contain several paleomagnetic results. Then there is a one-to-many association between them.

REFERENCE # ≪ ----------------≫ CONTINENT

M:N Association

Each reference may contain information about several continents but each continent may be referred to in several references (many-to-many association).

In paleomagnetism we tend to think of our data in a hierarchical way but, as already noted, it is at present more efficient to use a relational database system. Codd (1970) popularised the use of relations and tables as a way to model data. The relational data model uses the concept of a *relation* to represent what has previously been referred to as a *file*. A *relation* is merely a two-dimensional table containing columns and rows. Each column contains values about the same *attribute*. A *tuple* is a collection of values that comprise one row of a relation. At least one of the attributes in a table (or sometimes several in combination) uniquely defines each row in the table and is referred to as the *primary key*. There are various rules to be followed in defining a relation but these will not be discussed here.

Codd (1970) and others have shown that relations are actually formal operations

Surveys in Geophysics **12**: 317–491, 1991.
© 1991 *Kluwer Academic Publishers. Printed in the Netherlands.*

on mathematical sets. Furthermore, most data processing operations, (e.g. the printing of selected records and finding of related records) can also be represented by mathematical operations on relations. The results of these mathematical operations can be proved to have certain properties. A collection of such operations has been shown to result in databases with desirable logical and maintenance properties. This mathematical elegance and visual simplicity has made the relational data model one of the driving forces in the information systems field. A readable reference on these matters is McFadden and Hoffer (1985).

The technique of *NORMALISATION* provides a foundation for logical database design. Normalisation is the analysis of the associations between attributes and has the purpose of reducing what may be a complex user view of the data to a set of small, stable data structures. Experience clearly shows that normalised data structures are more flexible, stable and easier to maintain than unnormalised ones. Several basic steps have to be undertaken in the normalisation process. There are at least five normal forms but usually the first three normal forms are sufficient for most databases. This will certainly be the case for most paleomagnetic databases where the degree of complexity is relatively low.

2. Choice of Database Management System

The Relational Database Management System Language SQL (Structured Query Language) was developed by IBM in 1977. However it was first implemented by ORACLE in 1979. Subsequently IBM implemented SQL/DS in 1982 and DB2 in 1984. Since 1979 ORACLE has been at the forefront of the RDBMS market and introduced its own utility SQL*Plus that includes editing, formatting and special system commands. In 1988 the International Standards Organisation recognised SQL as the official international language for RDBMS. Furthermore ORACLE has now developed the first PC version with built in SQL*Plus and all the features of the mainframe version.

Although mainframe versions can implement the complete range of RDBMS functions, there are many PC systems that have been developed from the perspective of limited memory availability. The result of this is that many functions are often missing in PC-RDBMS products. For example, there is frequently no restart and recovery support and no data security procedure. Often no transaction log or audit trail is maintained. Major problems arise with the maintenance of integrity, consistency, security and with normalisation. These factors are crucial in designing and setting up a database that is robust for use by the international community.

At the time the GPMDB was first proposed ORACLE was the only RDBMS with mainframe and PC versions that had the same characteristic. It is now a simple matter to expand the working memory of a PC (AT or above) to 2 or more Megabytes. With a built in 20 Mb, or preferably 40 Hb, hard disk it is now a simple matter to implement PC ORACLE with the full range of capabilities for use with a paleomagnetic database.

ORACLE had several characteristics that were not found in other RDBMS available at the time it was chosen for the GPMDB. For example a distinction is made between zero, blank space and nothing (called NULL values). Where an attribute in a relation may contain NULL values (i.e. there is no entry at all) then the system *does not store anything*. If a field is declared to be 60 characters long and if only 20 characters of information are required then only 20 are stored. This provides great flexibility in design. Without overwhelming the storage space, fields may be declared to be fairly large to cover those odd occasions when more space is needed. The uniqueness of primary keys in each record can be checked automatically on insertion. A system of *TRIGGERS* is available for insertion of data. With good design these can be a powerful tool for automatic checking of the validity of data on insertion.

Many of the leading Earth Science organisations and Universities in the western world now use ORACLE. The versatility of having PC and mainframe compatible versions and its great flexibility was such that the Czech Workshop in 1988 over-whelmingly endorsed the development of the GPMDB through ORACLE. This does not mean that the Database cannot be used under other RDBMS, but what it does mean is that under systems other than ORACLE its value may be greatly degraded. A readable reference about ORACLE is Perry and Lateer (1989).

One of the major advantages of RDBMSs is their ability to join together, on the basis of attributes, different tables in order to provide answers to very sophisticated queries. It should be noted that most simple database systems such as dBASE were developed around the concept that a 'database' is a single table. Until the advent of dBASE IV it was only possible to join two tables together in any one query. A full relational database however is a collection of tables that can be manipulated and joined using the SQL language. Although dBASE IV now im-plements SQL as an addendum to its own language the full power of relational operators is still not available. Our view is that ORACLE is a system that looks ahead to future developments in the RDBMS field. However in consideration of those who feel they cannot afford the more complex system or who already possess a simple database system, we have developed two versions of the GPMDB – one for ORACLE and a 'compact database' called ABASE for use with simpler systems as described below.

3. Design of the Global Paleomagnetic Database

3.1. ORACLE DESIGN

In designing a relational database for paleomagnetic directions and poles the first step is to identify the basic entities involved. We have followed the usual form in

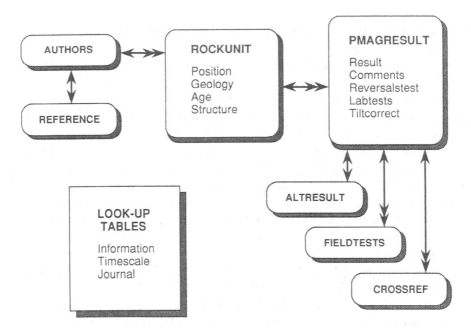

Fig. 1.   ORACLE version of the Global Paleomagnetic Database. Tables are linked as either one-to-one (single headed arrows) or one-to-many (double headed arrows) associations.

which paleomagnetists present their data. This form follows the one that was used when the Pole Lists of the Geophysical Journal were publised on a regular basis. In this format a journal reference gives sampling details for a rockunit from which one or more magnetisation components or paleomagnetic results are derived. Thus there are only three basic entities. However, the process of normalisation demands that where there are many-to-many associations between entities, these must first be removed so that only one-to-one or one-to-many associations remain. The result of this process is to divide some entities into several entities or else to create some artificial entity as an intermediate step.

A design that takes the above considerations into account is shown in Figure 1. There are seven relations (tables) that are linked as one-to-many or one-to-one associations. We have maintained an AUTHORS table separated from the corresponding REFERENCE table largely because this was the way in which the database was first developed historically. However, this separation is not necessary and will most probably be eliminated when the first update of the database is made after 1992. Also we have found it convenient to link AUTHORS to a special table called REMARKS that is only available in the SQL*Forms method of querying the database (see Section IV).

Each set of AUTHORS and associated REFERENCE will relate to one or more ROCKUNIT. The ROCKUNIT table contains details of the sampling for each paleomagnetic study. This table is not one that describes the geology of a particular ROCKNAME, but rather the specific geological parameters related to the study in question. The important parameters for the paleomagnetist are the position and time interval of the rocks that were sampled. Thus the same ROCK-NAME may appear in several records but the sampling details will be different in each case. Because of this there are no duplicated records and both AUTHORS and REFERENCE are in one-to-many association with ROCKUNIT. Each ROCKUNIT may have several results (PMAGRESULT) associated with it. Continuing down the chain then each PMAGRESULT may have several FIELD-TESTS associated with it and so on.

The design chosen makes the structure of the database more readily understood by paleomagnetic users. It provides a set of tables that are linked in a sequence that most workers in the subject follow and understand. This sequence was particularly useful when it came to designing the custombuilt Form under SQL*Forms (see Section IV). Users page down the Form in a way that follows the traditional manner in which data have been presented.

Each relation in the design may be considered to be made up of several sub-tables as shown within each entity block in Figure 1. Each of these is in one-to-one with the basic information for that entity and it is not necessary to separate out one-to-one associations into separate tables. Apart from separating AUTHORS and REFERENCE as mentioned above, we have also created a separate table called ALTRESULT. Here the main table PMAGRESULT always gives the Pole Position as derived from Declination and Inclination and Site coordinates. However, if the result has been derived from the mean of site VGPs, details are then given in ALTRESULT. A separate set of 'look-up' tables provides subsidiary information.

Table I provides a list of the relations and attributes with the primary keys in each relation underlined. This shows that the primary keys of each relation have been progressively incorporated as part of succeeding relations. These have the purpose of carrying out joins between tables as described under in Section V.3.3 (Selecting Data from More than One Table – the Join clause).

Under ORACLE the GPMDB is secure. Users have only been granted select privileges on the tables and have no capability to change entries. Thus even if entries contain errors or spelling or other mistakes these must be reported for update when the next version is released. As is explained in Section V.6, users can copy the tables or parts of the tables into their own personal database with access restricted to themselves alone. These personal tables could of course be changed or altered in any way that is desired. The advantage of the security of the ORACLE database tables is that users could quote that a particular data set was derived through a stated SQL query of the ORACLE database. This query can then be repeated by anyone with the same version of the ORACLE database.

3.2. COMPACT DATABASE – ABASE

To assist those who wish to stick with a simpler database system we have also prepared a set of three basic tables as a set of ASCII files (Figure 2). This set of

TABLE I

Global Paleomagnetic Database – Relations and Attributes (Primary keys are underlined)

AUTHORS	(REFNO, AUTHORS)
REFERENCE	(REFNO, YEAR, JOURNAL, VOLUME, PAGES, TITLE)
ROCKUNIT	(REFNO, ROCKUNITNO, ROCKNAME, PLACE, CONTINENT, TERRANE, RLAT, RLONG, ROCKTYPE, STRATA, STRATAGE, LATSPREAD, LOWAGE, HIGHAGE, METHOD, ISOTOPEDATA, STRUCTURE)
PMAGRESULT	(ROCKUNITNO, RESULTNO, COMPONENT, LOMAGAGE, HIMAGAGE, TESTS, TILT, SLAT, SLONG, B, N, DEC, INC, KD, ED95, PLAT, PLONG, PTYPE, DP, DM, NOREVERSED, ANTIPODAL, N_NORM, D_NORM, I_NORM, K_NORM, ED_NORM, N_REV, D_REV, I_REV K_REV, ED_REV, DEMAGCODE, TREATMENT, LABDETAILS, ROCKMAG, N_TILT, D_UNCOR, I_UNCOR, K1, ED1, D_COR, I_COR, K2, ED2, COMMENTS)
ALTRESULT	(RESULTNO, APLAT, APLONG, KP, EP95)
FIELDTESTS	(RESULTNO, TESTTYPE, PARAMETERS, SIGNIFICANCE)
CROSSREF	(RESULTNO, CATNO)

LOOK-UP TABLES

INFORMATION	(SYMBOL, EXPLANATION)
TIMESCALE	(PERIOD, EPOCH, AGE, TOP, BASE)
JOURNAL	(ABBREVIATION, FULLNAME)

EXPLANATORY NOTES

1. AUTHORS: AUTHORS are listed according to a standard format as Author1,A.B., Author 2,A.B. with a space between each name.
2. REFERENCE: Book titles are included under JOURNAL and TITLE is given in full. The standard format for the most common journal names is listed in the table JOURNAL. Note that VOLUME is an integer and does not contain series letters.
3. ROCKUNIT: This relation can be considered as being made up of four sub-tables that for the purposes of simplicity of storage are placed together in the same relation. These are:
   (a) POSITION (ROCKNAME, PLACE, CONTINENT, TERRANE, RLAT, RLONG) Although several Rockunits may have been studied by different authors, each Study will be given a separate ROCKUNITNO to avoid problems with normalisation.
   PLACE always contains the name of a country that can be used for search purposes.
   CONTINENT lists one of 12 present-day continents including each of the main oceans. This enables searches of various parts of the globe easier. The continent name has an initial capital followed by lower case (e.g. North America, Africa).
   TERRANE lists a terrane name where this has been mentioned by the author or where the terrane is obvious to the compiler. However these terranes are not intended for search purposes, rather the idea is that the column can be copied with a user's personal database and names then inserted according to the need.
   RLAT, RLONG are the latitude and longitude of the Rockunit. The convention for RLONG is the range -180.0 to $+180.0$. Generally this refers to the mean sampling site, but where several results are recorded from the same unit SLAT, SLONG in the result table (PMAGRESULT) will not necessarily be the same for each of the entries.
   (b) GEOLOGY (ROCKTYPE, STRATA, STRATAGE, LATSPREAD)
   ROCKTYPE contains four possible Keywords that can be used for searching – extrusives, intrusives, sediments, metamorphics. These Keywords can be followed by others such as limestones, redbeds or basalts, andesites or gabbro, granite etc.
   STRATA usually provides more detailed stratigraphic information, for example "Visean".
   STRATAGE uses the common symbols for geologic periods followed by U,M,L to indicate Upper, Middle and Lower. These symbols are listed in the table TIMESCALE.

TABLE I. *(continued)*

(c) AGE (LOWAGE, HIGHAGE, METHOD, ISOTOPEDATA)
LOWAGE, HIGHAGE gives the estimated lower and upper limit numerically of the age of the rockunit. This may simply be a translation of the STRATAGE from the 1989 Geologic Time Scale, or may be more restricted according to the information contained under STRATA, or may be based solely on isotopic information given under ISOTOPEDATA below.
METHOD gives the basis upon which the LOWAGE and HIGHAGE values have been derived (e.g. fossils, Rb-Sr). Where no details have been given by the author as to how the age has been determined we use the general term stratigraphic.
ISOTOPEDATA lists any isotopic age information whether or not it is used to determine the LOWAGE and HIGHAGE.

(d) STRUCTURE. Comments – usually a range of strikes and dips or the age at which folding or metamorphism took place.

4. PMAGRESULT: This rather large relation can be considered as being made up of the following sub-tables:

(a) RESULT (COMPONENT, LOMAGAGE, HIMAGAGE, TESTS, TILT, SLAT, SLONG, B, N, DEC, INC, KD, ED95, PLAT, PLONG, PTYPE, DP, DM, NOREVERSED).
COMPONENT gives the name (or letter) given by the author where several components are identified.
LOMAGAGE and HIMAGAGE refer to the lower and upper best estimates of the magnetic age of the magnetization component, according to the author. Where necessary an explanation appears in COMMENTS.
TESTS are designated by the following symbols:
   F = Fold Test, F* = Synfold Test, Fs = Fold Test + strain removal
   C = Baked Contact Test, C* = Inverse Contact Test
   G = Conglomerate Test, G* = Intraformational Conglomerate Test
   U = Uncomformity Test (after Kirschvink)
   R = Reversals Test (Ra, Rb, Rc or Rai, Rbi, Rci)
   M = Rock Magnetic Tests
   N = No Tests
Where appropriate these are followed by + or − to indicate the test was positive or negative or by o to indicate it was inconclusive. For the Reversals Test the designation Ra, Rb, Rc is used following McFadden and McElhinny (1990). Rai, Rbi, Rci are used to indicate the isolated observation test. For the fieldtests details will be given under FIELDTESTS. Reversals data appear in REVERSALTEST below, and Rock Magnetic data appear under LABTESTS (ROCKMAG).
TILT will indicate the percentage tilt correction applied: 0 for no correction or 100 for full correction would be usual, but in cases of synfolding magnetization some intermediate value would be listed. Further details appear under TILTCORRECT below.
SLAT, SLONG is the site position for calculating the pole position and SLONG lies in the range −180.0 to +180.0.
DEC, INC, KD, ED95 are the usual directional parameters.
PLAT, PLONG give the pole position with PLONG given as 0 to 360 degrees by convention. Note difference for RLONG, SLONG.
PTYPE indicates how the pole position is calculated.
D = Pole calculated from mean DEC, INC and V = Pole calculated from mean VGP (details given under ALTRESULT). In the case of PTYPE = V, then DP = DM = EP95. NOREVERSED give the percentage of reversed data.

(b) REVERSALTEST (ANTIPODAL, N_NORM, D_NORM, I_NORM, K_NORM, ED_NORM, N_REV, D_REV, I_REV, K_REV, ED_REV).
ANTIPODAL gives the angle between N and R results according to NORM or REV data. Otherwise the usual directional parameters are given.

TABLE I. (continued)

(c) LABTESTS (DEMAGCODE, TREATMENT, LABDETAILS, ROCKMAG).
  DEMAGCODE summarises the laboratory and analytical procedures as:
    0 = No demagnetization
    1 = Only pilot demagnetization on some samples
    2 = All samples treated, blanket treatment only
    3 = Stereonets with J/Jo, or vector plots provided
    4 = Principal Component Analysis (PCA) plus vector or stereoplots and J/Jo
    5 = PCA plus vector plots plus multiple treatments that are successfully isolating vectors
      (e.g. AF and Thermal, or Thermal and Chemical etc).
  TREATMENT uses symbols A = AF, T = Thermal, H = Chemical and N = No treatment.
  LABDETAILS provides other information such as maximum AF field or Temp used for
  pilot test purposes.
  ROCKMAG summarises information on any basic Rock Magnetic studies with the five most
  common types being:
    OP (opaques from reflected light – MAGN = magnetite, MAGH = maghemite, TM =
      titanomagnetitie, TMH = titanomaghemite, ILM = ilmenite, HEM = hematite,
      THEM = titanohematite, GEOTH = geothite, PYRR = pyrrhotite, RUT = rutile,
      MART = martite)
    Js-T (Curie Temps)
    IRM (IRM acquisition with saturation field)
    SUSC (Susceptibility)
    AN (Anisotropy).
(d) TILTCORRECT (N_TILT, D_UNCOR, I_UNCOR, K1, ED1, D_COR, I_COR, K2,
  ED2) provides the data before and after tilt correction for N_TILT measurements.
(e) COMMENTS provides other information including details of the origin of MAGAGE.
  Reference to previous studies is given.
5. ALTRESULT (APLAT, APLONG, KP, EP95)
  This table provides the Pole Position and statistics when derived from the mean VGP. If this
  calculation is the only one given by the author then APLAT = PLAT and APLONG = PLONG
  and DP = DM = EP95 in the PMAGRESULT relation.
6. FIELDTESTS (TESTTYPE, PARAMETERS, SIGNIFICANCE)
  TESTTYPE uses the symbols F, F*, Fs, C, C*, G, G*, U as under TESTS.
  PARAMETERS contains a commentary and any data related to the particular test.
  SIGNIFICANCE indicates whether the test was positive or negative or inconclusive with
  calculated probability where possible.
7. CROSSREF: CATNO gives the equivalent result in other catalogues such as GJ (Pole Lists),
  OT (Ottawa), SU (Soviet). The style used in GJ 15.256, OT 030056 etc.
8. INFORMATION provides an explanation of the symbols used in various columns of the
  database.
9. TIMESCALE gives a listing of the parameters for the 1989 Geologic Time Scale of Harland *et
  al.* (1990).
10. JOURNAL gives a listing of the style used in REFERENCE for the most common journals.

ASCII files (ABASE) consists of three basic tables that could readily be used with dBASE or a similar Database Management System. Details are given in Table II. This takes into account that many of these systems cannot join more than two tables at any one time.

This set of tables is now similar to the type of information that used to appear in the Geophysical Journal Pole Lists. It should be much easier to manipulate using dBASE or other systems than the complete set of ORACLE tables. The

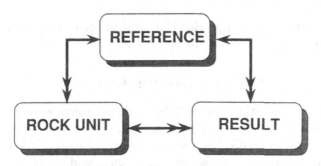

. Fig. 2.   The compact database (ABASE) consisting of three tables where any pair of tables may be
linked if required.

main advantage of these tables is that we have selected those columns that are in
almost all cases filled with some information so that the number of blank spaces
to be stored is minimised. Note that RESULT contains all three primary keys so
that it may be linked to either of the other two tables directly. Even so the details
given in Table II show that already even this so-called Compact Database is
quite large and we hope that, despite the restrictions imposed, MS-DOS without
extended memory will be able to cope with this set of tables. It is possible for
other sets of tables to be produced from ORACLE if required.

The tables provided for ABASE are not secure and can be altered and changed
at will. Users of this form of the database will not be in a position to claim that

TABLE II
Set of Non-ORACLE ASCII files: Compact Database – ABASE

REF	(REFNO, AUTHORS, YEAR, JOURNAL, VOLUME, PAGES, TITLE)
ROCKUNIT	(REFNO, ROCKUNITNO, ROCKNAME, PLACE, CONTINENT, ROCKTYPE, STRATAGE, LOWAGE, HIGHAGE, METHOD)
RESULT	(REFNO, ROCKUNITNO, RESULTNO, COMPONENT, LOMAGAGE, HIMAGAGE, TESTS, TILT, SLAT, SLONG, B, N, DEC, INC, KD, ED95, PLAT, PLONG, DP, DM, NOREVERSED, DEMAGCODE, TREATMENT, COMMENTS)

FILE FORMAT

1. Each column is enclosed in inverted commas and columns are spaced with a comma. E.g.
   "column1", "column2", ... , "columnN"
2. Each record is continuous and terminates with a line feed. The next record then starts on a new
   line. The length of each record or line is given in the details below with the size of each file being
   that in January 1991.

TABLE AND COLUMN SPECIFICATIONS

TABLE NAME	NO COLUMNS	NO ROWS	LINESIZE	SIZE
REF	7	1862	433	810 Kb
ROCKUNIT	10	3354	261	882 Kb
RESULT	24	3785	265	1010 Kb

certain queries of this data set provided information used for a particular analysis, since it cannot be guaranteed that the set has not been altered.

4. Aspects of Data Entry

The explanatory notes at the end of Table I provide some details of the various columns and how they have been filled. There are several points that need clarifying. We have tried to keep up with revisions to rockunit ages wherever possible but we cannot guarantee that things have not been missed. Thus ages given in the original papers may have been revised in the light of subsequent information. This is usually explained or cross-referenced. Obviously the older the original reference the more likely this is to have taken place.

For ease of searching for data in different parts of the world we have specified that each rockunit belongs to some present-day continent. For this purpose we have specified the usual continents (including Greenland as a continent) and all the major oceans giving 12 possibilities in all. Thus it is a simple matter to request data from North America or the Pacific Ocean. A listing of these 'continents' can be obtained by querying the database as described in Section V.2.8 (Selecting Unique Rows from a Table) and is also given in Figure 26.

For the data listed in the Geophyslcal Journal Pole Lists 1 through 7, we have used combinations made by Irving (1964) that simplify these entries rather than presenting every detailed result that was initially used. Many of these earlier results are in any case of historical interest only nowadays. These results tend also to have fewer column entries in that no demagnetization studies or rock magnetic studies were made. Where recalculations have been made for data in the time range 0–5 Ma (especially in the case of lava sequences), we have followed the usual convention of omitting all VGPs wlth latitude less than 45 degrees.

The magnetic age of the magnetizations creates a problem. In each case we have retained the view expressed in the original paper, however the older data were subjected to somewhat less rigorous analysis than nowadays and it was usually assumed that the primary magnetization had been determined. It is often not made clear in earlier results whether tilt corrections were applied. If no mention is made of this we assume that none has been applied unless it is obvious that a tilt correction was made. This clearly may not be true.

Updates of data by an investigator are treated as though a new study has been made except where the original paper did not use cleaning techniques. In these cases we have not listed the original result (uncleaned) but make reference to it under the COMMENTS column.

It is worth commenting on the use of the Baked Contact Tests. The complete test requires that the baked rock and the baking rock have the same magnetization direction. In addition the unbaked rock must be shown to have a different magnetization to demonstrate that the agreement between baked and baking rocks is not due to some regional remagnetization. There are many studies in the literature where only the baked and baking rocks are studied and a positive baked contact test is claimed. Where no unbaked studies have been made we have designated the test indeterminate using the symbols 'Co'. When detailed studies are made of the stable magnetization of the unbaked rocks, then the baked contact test is evidence that the magnetization of these unbaked rocks has been stable at least since the time of baking. We have referred to this as the Inverse Contact Test (designated by 'C*').

II. ORACLE HARDWARE AND SOFTWARE

1. System Hardware and Software Recommendation for ORACLE RDBMS

1.1. FOR A FULLY IBM COMPATIBLE PERSONAL COMPUTER (PC)

Hardware Requirements

To install and run ORACLE for MS-DOS, your computer must meet these minimum requirements which are for Version 5.1C in 1990. These requirements are subject to change without notice in future versions.

- An IBM Personal Computer AT or above, or Oracle-certified 100% compatible computer capable of running Microsoft MS-DOS. We have successfully installed and run ORACLE on various non-certified computers around the world (including our own) but bear in mind that ORACLE gives no guarantee of success on non-certified machines.
- 640 K bytes of random-access memory (RAM)
- A minimum of 896 K bytes of additional EXTENDED memory – not expanded or enhanced expanded memory. 2.5 Megabytes are recommended by ORACLE if using application tools in Protected mode. For the GPMDB we RECOMMEND 2 extra Megabytes of EXTENDED memory over and above the 1 Megabyte which is supplied with your system. This makes a total of 3 Megabytes of RAM.
- One $5\frac{1}{4}$ inch, double-sided diskette drive.
- A fixed (hard) disk. To install ORACLE for MS-DOS a minimum of 8 Megabytes of hard disk space is required for the database and database software in addition to the disk space required by MS-DOS and the ORACLE tools. The ORACLE tools require up to an additional 12 Megabytes of disk space. Table III lists a complete breakdown of the RAM and hard disk space required by each ORACLE product.
- A colour or monochrome display (VDU, CRT) supported by MS-DOS.

Software Requirements

- Version 5.1C ORACLE for MS-DOS. We expect that later versions will be compatible with the database EXPORT and the customised programs and Forms we have developed. It is ORACLE's policy that implementations developed on earlier versions are supported by later versions.
- Microsoft's MS-DOS operating system, Version 3.1 or later configured for American ASCII.

TABLE III

Minimum memory requirements for both RAM and hard disk for the ORACLE products that might be used with the GPMDB

| Product | Executable | Total RAM | | | Hard Disk Space | | |
		Protected mode	Real mode		Protected mode	Real mode	Both modes
RDBMS	ORACLE	896K	N/A				
	SQLPME	10K	and	60K			
RDBMS Total		906K	and	60K	4900K	N/A	N/A
Utilities	IMP	150K	150K				
	EXP	150K	150K				
	SGI	75K	75K				
	EXPAND	125K	125K				
Utilities Total		500K	500K		1075K	1105K	2055K
SQL*Forms	SQLFORMS	800K	575K				
	RUNFORM	400K	400K				
	IAC	225K	225K				
	IAG	125K	125K				
Forms Total		1150K	1325K		2240K	2205K	3285K
SQL*Menu	SQLMENU	675K	325K				
	DMD	200K	200K				
	DMM	200K	200K				
Menu Total		1075K	725K		1425K	1015K	1920K
SQL*Plus	SQLPLUS	425K	425K		670K	675K	950K
SQL*Loader		350K	350K				
Tools Utilities		375K	375K		460K	460K	760K

1.2. FOR AN APPLE MACINTOSH PC

Hardware Requirements

- A MAC SE, or above. A MAC Plus capable of running System 6.0 and Finder 6.1 may also be used.
- 2 Megabytes of RAM
- A fixed (hard) disk with at least 5 Megabytes of free space available to install the basic programs and create the additional files needed for modestly-sized databases. See note above regarding the size of the GPMDB.
- A VDU

Software Requirements

- ORACLE for Macintosh
- It is more convenient to run ORACLE under the Macintosh Multi-Finder, but if you have only 2 Megabytes of RAM you may run ORACLE under the Finder.
- System 6.0 or later.
- Finder 6.1 or later.
- HyperCard Version 1.2, in order to run the ORACLE for Macintosh System Stack and demonstration stacks.
- Macintosh Programmer's Workshop (MPW), Version 2.0 or 3.0. If you wish to use all the features of the ORACLE for Macintosh SQL*Plus module, you will need MPW. MPW is a complete set of programming tools which includes a line-oriented command interpreter, a windowing editor, assembler, linker and text analysis tool. MPW will enable you to use the SQL command files we have developed for the GPMDB from within SQL*Plus. However all ORACLE database functions may be performed within HyperCard without MPW.

1.3. RECOMMENDATIONS

We recommend an IBM compatible PC as this will enable you to use the customised SQL*Forms Application provided with the GPMDB. Forms is a powerful part of the ORACLE RDBMS query utilities and are also readily ported between other IBM compatible PCs, minis, mainframes or via networks. The APPLE MACINTOSH cannot at present offer this facility although it supports a range of other ORACLE tools. HyperCard can be used to develop Form-like applications for MACINTOSH but as yet ORACLE does not support a Forms equivalent for MACINTOSH. Trial runs using the GPMDB with Macintosh's HyperCard on a Mac II indicate that it is significantly slower (up to an order of magnitude!) than SQL*Forms on MS-DOS machines in retrieving information from the database.

We recommend at least 40 Megabyte hard disk. The minimum disk space of 8 Megabytes includes approximately 3 Megabytes of database file space. This is *NOT* sufficient to store the GPMDB (years 1950–1988). The GPMDB currently requires about 8 Megabytes to store and query the database. To allow for future growth of the database beyond 1988, and increases in the size of other essential ORACLE operating files, we recommend at least 40 Megabyte hard disk partitioned as drive C: 30 or 32 Megabytes and drive D: 10 or 8 Megabytes. Some hard disk drives are supplied with software that allows the creation of partitions greater than 32 Megabytes. *Some of these drivers may cause problems with ORACLE* even if the database is less than 32 Megabytes. ORACLE therefore strongly recommends that hard disk partitions be kept to a maximum of 32 Megabytes.

We recommend the purchase of an advanced disk management package such as XTreePro, XTreeGold or Norton Utilities. These are easy to use and will greatly facilitate management of your MS-DOS directories and files. They have the advantage that they may readily be used from within the ORACLE software. XTreePro is *NOT* compatible with MS-DOS Version 4 if a SHELL command is used. Section III.2.2 (The AUTOEXEC.BAT File) discusses this more fully.

Norton Utilities includes very useful disk management, recovery and defragmentation utilities. The Norton Integrator (NI) organises all the Norton Utilities. The Speed Disk Utility packs the disk to eliminate file fragmentation, and prioritises directory and file placement. ORACLE's database files, which store the GPMDB, require large amounts of contiguous file space for efficient operation. When installing or updating the GPMDB the hard disk should be packed to ensure that the maximum contiguous file space possible is available.

2. ORACLE RDBMS Software Used By the GPMDB

The following information assumes that you have chosen an ORACLE certified IBM compatible PC using an MS-DOS operating system. Listed below, with brief descriptions, are the ORACLE products you will need to install to use the GPMDB.

2.1. A BRIEF DESCRIPTION OF THE ORACLE RDBMS

ORACLE is a terminate and stay resident (TSR) program (i.e. after shutting down ORACLE continues to occupy memory even though it is not running). If you attempt to start another large program you may receive error messages or encounter other difficulties. If this is the case ORACLE must be unloaded. See Section III.5.2 (Stopping the ORACLE RDBMS) for a description of how to do this.

The Database Administrator

If you will be the only user of the GPMDB on a PC then you must act as the Database Administrator (DBA) for your system. If there is more than one user of the GPMDB on your system, one person should be appointed to be the DBA. The DBA is responsible for configuring your PC system for ORACLE, installing ORACLE, loading the GPMDB and the custombuilt files, managing and maintaining the database. These duties are described in detail in Section III (Database Installation and Maintenance).

ORACLE Utilities

The ORACLE RDBMS includes facilities for maintaining the database, for performing administrative tasks such as backing up and restoring data and handling space allocation.

CCF – Creates MS-DOS files used to store database information

EXP – Exports data from the database to an encoded, very compressed format, MS-DOS file.

IMP – Imports data exported by EXP.

IOR – Initialises a database, starts and stops ORACLE.

SGI – Displays the size of the System Global Area.

ORACLE Tools

SQL*Plus   An interactive SQL interface application tool. The SQL*Plus command language is an extension of the SQL language. SQL*Plus can use SQL statements to retrieve data from the database via queries, to create and delete your own tables and views, and to act as a DBA. SQL*Plus enables queries to be stored, retrieved and edited, and output from queries to be formatted. After installation of ORACLE the DBA must, using SQL*Plus, issue a userID and password to grant each user logon and database resource privileges.

SQL*Forms   An interactive application program for designing and using Form-based applications. A Form is a screen display that resembles a printed form and it is an easy way to access data in a database. SQL*Forms can be used to enter data. This is how we entered the data into the GPMDB. It can also be used to display information retrieved from the database. We supply a custombuilt Form called PALEOMAG with the GPMDB for this purpose. After installation of SQL*Forms a user may logon and use the database using the same userID and password that was issued for logging into SQL* Plus.

SQL*Menu   A menu-generation program that allows the creation of full-feature end-user menus. A selection of tasks can be presented on one or more menus. A menu is a list of choices that either run software programs or call other menus. We supply a custombuilt menu called PMAG with the GPMDB. After installation of SQL*Menu the DBA will need to assign new user privileges to each user for the userId and password that was issued for logging into SQL*Plus.

Refer to Table III for the RAM and hard disk memory requirements for the above products. The GPMDB can be used without SQL*Menu but this tool makes using the database very user friendly. The GPMDB can also be used without SQL* Forms but the Form is a very convenient way of querying the database requiring minimum knowledge of SQL*Plus.

2.2. SQL*Plus under MS-DOS

To use SQL*Plus with the GPMDB after ORACLE has been installed and the GPMDB and custombuilt files loaded, enter the following commands:

C:\ >GORACLE   loads the ORACLE RDBMS
C:\X >SQLPLUS userID/password   starts SQL*Plus, if userID/ password are not
                                supplied you will be prompted for them
SQL> Enter SQL and/or SQL*Plus commands
SQL> EXIT   exit SQLPLUS and return to MS-DOS
C:\X >SHUTORA   shuts down the ORACLE RDBMS and re-
                                moves it from RAM

X represents your current directory and this will usually be the database work
directory, C:\PALEOMAG.

*Using SQL*Plus in Protected Mode*

SQL*Plus may be installed in REAL(R), PROTECTED(P) or BOTH(B) modes.
R and P modes require the same amount of hard disk space and RAM. However
the advantage of installing in P mode is that real memory space below MS-DOS's
640 K limit (within the 1 Megabyte of system-supplied RAM) is freed to allow the
use of networking software, TSR programs, an editor, a word processor or another
utility or program.

The LOGIN.SQL File

The LOGIN.SQL file is an ASCII file containing SQL statements and/or SQL*
Plus commands. When SQL*Plus is started it looks for the LOGIN.SQL file in
your current directory and executes the commands contained in it. The LOG-
IN.SQL file is not essential to the operation of SQL*Plus and if it is not found
SQL*Plus will start normally without giving an error message. However, the
LOGIN.SQL file is a convenient means of 'customising' SQL*Plus to fit your
particular needs. See the discussion below in 'Using an MS-DOS Editor' for an
example of this.

The ORACLE installation procedure places a standard LOGIN.SQL and a
non-standard LOGIN.NEW file in the ORACLE home directory (C:\ORACLE5).
SQL*Plus will use LOGIN.SQL as the default file on starting up if you have no
LOGIN.SQL file in your work directory. Section III.4.3 (Loading the Custombuilt
Files) describes how to place a copy of the appropriate LOGIN.SQL file in your
work directory.

Executing MS-DOS Commands

Any MS-DOS command can be executed without leaving SQL*Plus by prefixing
the command with the SQL*Plus command HOST or $. For example if you wish
to see the contents of your current directory then enter either of the following
commands:

SQL> HOST DIR

SQL> $ DIR

After the command has been executed, control returns to SQL*Plus and the standard SQL*Plus prompt is displayed. If HOST, with no command is entered, then you exit to MS-DOS. To return to SQL*Plus, enter EXIT.

SQL> HOST
C:/PALEOMAG >MS-DOS commands
C:/PALEOMAG >EXIT
SQL>

NOTE: do not attempt to logon to the database a second time using the HOST command. For example do NOT enter:

SQL> HOST ORACLE
SQL> HOST SQLPLUS

Using an MS-DOS Editor

EDIT is a SQL*Plus command which by default accesses EDLIN, the standard MS-DOS editor. However EDIT can be redefined to invoke almost any MS-DOS editor, or word processor from within SQL*Plus. To redefine the EDIT command enter:

SQL> DEFINE _EDITOR = 'xx'

where 'xx' is the command you would normally use to invoke your favourite editor or word processor.

To automatically assign your chosen editing software to EDIT each time you log into SQLPlus place the DEFINE _EDITOR command in the LOGIN.SQL file in your work directory. If you use a word processor, make sure the output text is a normal ASCII file using the mode that does not automatically include formatting and printing commands in the output file.

If you wish to edit the contents of the current buffer enter EDIT with no parameters:

SQL> EDIT

After you save the edited text, ORACLE places it back in the current buffer. To edit the contents of a file enter:

SQL> EDIT filename[.extension]

Specifying the file extension is optional. If it is not given EDIT assumes a default extension of .SQL.

Spooling to the Printer

The SPOOL command may be used to create an MS-DOS file, with the default extension .LST, containing the output from a query, or the contents of the SQL or some other buffer. To terminate spooling issue the SPOOL OFF command.

> SQL> SPOOL filename;
> SQL> SQL query;
> SQL> SPOOL OFF;

The contents of the SPOOL file may be edited or printed. The easiest (but not the only) way to print a SPOOL file is via the HOST command.

> SQL> HOST COPY filename.LST LPT1;

Where LPT1 is the printer port name. Your printer port name may be PRN, LPT1, LPT2, ... or COM1, COM2, ... or wherever you have your printer.

The MODE command may be used to redirect output from the LPT port to a COM port as in the following example:

> C:\ >MODE COM1:

> C:\ >MODE LPT1: = COM1:

The SPOOL command can be used to direct output directly to the printer. For example:

> SQL> SPOOL LPT1;
> SQL> SQL query;
> SQL> SPOOL OFF;

Limitations Under MS-DOS

When running SQL*Plus in R mode with TSR and other memory resident programs loaded, it may not be possible to edit SQL statements in the SQL buffer because there is insufficient memory (RAM). Also, it may not be possible to use the HOST command to exit to the operating system (MS-DOS). If this occurs, EXIT from SQL*Plus, remove the TSR and other memory resident programs from RAM, then start SQL*Plus again. When running SQL*Plus in P mode this restriction should not be encountered.

2.3. SQL*FORMS UNDER MS-DOS

SQL*Forms comprises the Forms Designer (SQLFORMS) and the Forms Application (RUNFORM). When a Form is first designed using SQLFORMS, the Form code is generated within the database in 'database format'. Then a file in intermediate format is generated and stored as an MS-DOS file called formname.-INP. The .INP file is then used to generate the binary code that RUNFORM requires to access the Form. This code is stored in an MS-DOS file called formname.FRM.

SQL*Forms includes the following Applications:

IAC – the Interactive Application Convertor. IAC converts Forms from .INP format to database format. The LOAD option in SQLFORMS runs IAC on an existing .INP file.

IAG – the Interactive Application Generator. IAG reads an .INP file and generates an .FRM file. The GENERATE option in SQLFORMS runs IAG.

IAP – the Interactive Application Processor. IAP or RUNFORM reads a Form from an .FRM file and runs it. The RUN option in SQLFORMS runs IAP.

If you do not intend to design your own Forms you will need to use only the IAP, RUNFORM. To run the GPMDB Form, PALEOMAG enter the following commands:

C:\ >GORACLE	loads the ORACLE RDBMS
C:\X >RUNFORM PALEOMAG	loads the Form PALEOMAG
userID/password	[Esc] exits the Form and returns to
(if no userID/password are given	MS-DOS
you will be prompted for them)	
C:\X >SHUTORA	shuts the ORACLE RDBMS down and removes it from RAM

X represents your current directory, usually the database work directory, C:\PALEOMAG.

*Installing the SQL*Forms Extended Data Dictionary*

When installing SQL*Forms for the first time, you can either install the Extended Data Dictionary tables automatically during the SQL*Forms installation, or you may create the tables manually after installation.

If you are sure you want to design your own Form Applications then answer Y (Yes) when prompted to have these tables installed automatically durinq ORA-

CLE installation. This is described in Section III.3.3 (Step by Step ORACLE RDBMS Installation Instructions.

If subsequently you decide that you wish to design your own Form applications then create the tables manually by running the file FORMINS.BAT from the C:\ORACLE5\BIN directory:

```
C:\ >CD C:\ORACLE5\BIN
C:\ORACLE5\BIN >FORMINS SYSTEM password
```

The FORMINS.BAT file runs the files: IADTABLE.SQL
                                    SETUPIAD.SQL

Dropping the SQL*Forms Extended Data Dictionary

SQL*Forms Extended Data Dictionary tables take up considerable hard dlsk space. If you loaded the Forms Designer during ORACLE installation and have subsequently decided that you do not wish to design your own Form applications you can drop the Extended Data Dictionary tables to release hard disk space. To drop the SQL*Forms Extended Data Dictionary, run the file IADREM.SQL from the C:\ORACLE5\DBS directory within SQL*Plus as follows:

```
C:\ >CD C:\ORACLE5\DBS
C:\ORACLE5\DBS >SQLPLUS SYSTEM/password
SQL> START IADREM
SQL> EXIT
```

Using SQL*Forms in Real Mode

If you wish to design your own Form applications do not load SQL*Forms in R mode as there will not be access to sufficient RAM. If your display has different characteristics to the type on which the custombuilt Form was developed you may need to make use of the Forms Designer software in R mode. This is described in the Section III.5.3 (Loading the Custombuilt Files). It is therefore useful to understand the difference between operating the designer in R and P mode. In R mode, RUNFORM operation must be run separately from the Forms Designer (SQLFORMS). This is due to the MS-DOS 640 K RAM limit. In order to generate the binary .FRM file that RUNFORM requires, you must exit the SQL*Forms designer and run IAG.

```
C:\PALEOMAG >IAG [path\]formname -to
```

where
- path is the directory where the Form is stored. If this is already your current directory then it need not be specified.

- formname is the name of the Form.
- -t IAG option that suppresses displaying the process on the screen.
- -o IAG option that prevents the .INP file being over-written.

You can now run a R mode version of the Form by issuing the RUNFORM command from the MS-DOS command line:

C:\PALEOMAG >RUNFORM [path\]formname userID/password

where
- path is the directory where the Form is stored.
- formname is the name of the Form.
- userID/password

If you do not supply a userID/password you will be prompted for them.

Using SQL*Forms in Protected Mode

When SQL*Forms is installed in P mode larger Forms can be designed and run. Some restrictions do still exist (refer to the your 'ORACLE for MS-DOS' documentation).

In P mode, RUN (the equivalent of RUNFORM) can be run from within SQLFORMS. Once a Form design is finalised the P mode version of the Form can be run from MS-DOS using RUNFORM.

2.4. SQL*MENU UNDER MS-DOS

To use SQL*Menu with the GPMDB enter the following commands:

C:\ >LOADORA	loads the ORACLE RDBMS, starts SQL* Menu and prompts you for your userID/ password
C:\ >SQLMENU userID/ password	starts SQL*Menu if ORACLE is already loaded, if you do not supply userID/ password you will be prompted for them

SQL*Menu invokes RUNFORM in order to process Forms that are needed for various SQL*Menu functions.

SQL*Menu in Real Mode

Due to the 640 K limit on RAM set by MS-DOS, neither SQL*Plus nor SQL* Forms can be linked to the R mode version of SQL*Menu.

2.5. THE .CRT FILE

Several .CRT files are supplied with the ORACLE RDBMS for use with different types of graphics adaptor cards and display types. A .CRT file contains a CRT definition that 'maps' PC keys to ORACLE functions. The task performed by any PC key you press is determined by the way that key is defined in the .CRT file. The .CRT file is selected when the ORACLE RDBMS is installed. The ORACLE Application Tools that need CRT definitions are SQL*Forms and SQL*Menu.

The terminal and keyboard characteristics defined by the .CRT file include:

- the type of display used (monochrome, CGA, EGA or VGA)
- the display mode (of such things as background colour, highlight colour, colour of regular text, colour of messages)
- key stroke sequences used to invoke various commands in the ORACLE tools that require it

The ORACLE tools that require a .CRT file can either use the one stored in the DEFAULT.CRT file or the one specified as an option when startinq the tool. The ORACLE installation procedure creates the DEFAULT.CRT file by copying either BIOS.CRT or the .CRT file you specify. The contents of the DE-FAULT.CRT file can be changed at any time after installation by copying a different .CRT flle into the DEFAULT.CRT file. All of the supplied .CRT files and the current DEFAULT.CRT file are stored in the C:\ORACLE5\DBS directory. The characteristics and descriptions of the available .CRT files are listed in your ORACLE for MS-DOS documentation. We find the colours used by BIOSBLUE.CRT (blue background) easier on the eyes than BIOS.CRT (black background) and the function key stroke assignments are the same.

III. DATABASE INSTALLATION AND MAINTENANCE

1. The Role of the Database Administrator

It is advisable for each PC system using the ORACLE RDBMS that a single person be appointed as the Database Administrator (DBA). The DBA should be responsible for managing and maintaining the GPMDB. The DBA should have sole authority to install or re-initialise the ORACLE RDBMS, expand the database, load, update and backup the GPMDB, and grant userID's (often with different privileges). This chapter is directed at your DBA.

On a PC system the initial DBA tasks are:

- Preparing the PC for loading ORACLE
- Loading the ORACLE RDBMS
- Changing the SYS and SYSTEM passwords
- Setting up userID's and passwords
- IMPorting the GPMDB
- Creating directories and loading the custombuilt files supplied with the GPMDB

2. Preparing Your PC for the Installation of the ORACLE RDBMS

2.1. ORACLE RDBMS for MS-DOS and extended memory

Although MS-DOS limits real memory space (RAM) available to run programs to 640K, ORACLE extends this limit with the SQL Protected Mode Executive (SQLPME). SQLPME allows the RDBMS and any of its optional application tools to address up to 15 Megabytes of additional (*EXTENDED*) memory above the 640K limit set by MS-DOS. ORACLE requires an 80286, 80386 or 80486 processor-based IBM, or 100% compatible, to run in extended memory ('protected mode'). ORACLE runs in the protected mode of these processors by using their ability to directly address memory above 640K. For more information on SQLPME refer to 'SQLPME' in this section.

Most sophisticated applications that use extra memory use *EXPANDED* memory but ORACLE uses *EXTENDED* memory. These two are different. Most, if not all, popular memory cards can be configured as either *EXTENDED* or *EXPANDED*. Therefore you must make sure that your memory card is configured correctly.

Checking Extended Memory

Make sure before you first install the ORACLE Database for MS-DOS on your PC that there is enough *EXTENDED* memory. To do this power on your system and note the maximum number that the memory check counts. Only base memory

TABLE IV

Limitations of R versus P mode for the ORACLE application tools that may be loaded for use with the GPMDB

Application tool	Real mode	Protected mode
SQL*Plus	Possibly cannot use EDIT and HOST commands	Can use both EDIT and HOST commands
SQL*Forms	Limited design capacity	Can design larger Forms applications
SQL*Menu	Cannot link to R or P mode SQL*Plus or SQL*Forms	Can link to R and P mode SQL*Plus and SQL*Forms

and memory configured as *EXTENDED* are counted at this time. Memory configured as *EXPANDED* will be displayed later after it is loaded.

ORACLE requires a minimum of 1536K of usable memory (640K of real memory plus 896K of expanded memory), to proceed with installation in R mode only. Some machines also count system memory for MS-DOS, in which case the memory counter should reach at least 1920K. If you have followed our recommendations in Section II.1 (System Hardware and Software Requirements for ORACLE RDBMS) then 2688K (640K of real memory and $2 \times 1024K$ of *EXTENDED* memory) should be counted and installation done in P mode only. If the memory count does not reach the correct number then consult the memory board's instruction manual for information on the correct configuration and installation of the board.

If a memory check or a SETUP error occurs (these errors are issued by the MS-DOS system), but the memory check procedure counts the correct number of kilobytes, then run the system's SETUP to acknowledge the *EXTENDED* memory.

SQLPME

SQLPME is the ORACLE enhancement to MS-DOS that allows the RDBMS and its application tools to run in extended memory, or 'protected mode'. The ORACLE RDMS for MS-DOS automatically installs SQLPME for the ORACLE kernel in P mode. The optional application tools may be installed for use in R, P or B modes.

Mode of Installation of the Application Tools

Whether the application tools should be installed in R, P or B modes depends on the configuration of your system and whether you plan to design Form and/or Menu Applications. Electing to install in R mode imposes some restrictions on the use of application tools. Table IV sumarises these restrictions. For maximum flexibility we recommend that all tools be installed in P mode.

Real Mode   If you have only 1 extra Megabyte of *EXTENDED* memory
                   that is to be used to run the RDBMS, then it will be necessary
                   to use R mode to run the application tools. However, this will
                   restrict the disk space available to run the SQL*Forms Designer
                   and limit you to the design of small Forms Applications. Note
                   that this option requires the least hard disk storage for ORA-
                   CLE and *EXTENDED* memory at run time. Initially, it may
                   be possible to manage with 20 Megabytes of hard disk but this
                   will not allow for future growth of the database.

Protected Mode   If you have adequate *EXTENDED* memory (from 2 extra
                   Megabytes that we recommend up to a maximum of 15 Mega-
                   bytes), you should install all tools in P mode to obtain the
                   maximum flexibility and the enhanced capability that this offers
                   to run the application tools. If you plan to design Forms and/or
                   Menu applications, you must install ORACLE in P mode in
                   order to have access to sufficient memory.

Both Modes   If you have a limited amount of *EXTENDED* memory and
                   want the option to use some application tools in R mode and
                   others in P mode then install in B mode. The disadvantage of
                   this mode is that the PATH command in the AUTOEXEC.
                   BAT file has to be restructured or re-issued to ensure the
                   correct binary code is accessed for each mode. See the dis-
                   cussion below in 'The AUTOEXEC.BAT File'. Advanced MS-
                   DOS users will know how to overcome this problem by the use
                   of .BAT files. Note that this option requires the most hard disk
                   storage for ORACLE.

2.2. SETTING UP MS-DOS FOR ORACLE

For simplicity and clarity we assume that you have followed our recommendations
and purchased 2 Megabytes of extra RAM configured as *EXTENDED* memory;
that your hard disk is 40 Megabytes partitioned so that disk C: is 32 Megabytes
and disk D: is 8 Megabytes; that disk drive C: is where you boot MS-DOS; that
drive A: contains your $5\frac{1}{4}$ inch floppy diskette drive; and that you will follow the
ORACLE default installation procedure that installs and runs the RDBMS on the
C: disk. Note that you will require a minimum disk partition of 20 Megabytes to
load the ORACLE RDBMS and the GPMDB. As discussed previously, ORACLE
recommends that the maximum disk partition should be 32 Megabytes to ensure
there are no problems encountered in running ORACLE. Because of limitations
on the number of files that may be stored in the root directory (C:\) it is good
practice to store MS-DOS files in their own directory (C:\DOS). The only files
that should be placed in the root directory are discussed below. If your system

```
DEVICE=ANSI.SYS
FILES=40
BUFFERS=20
BREAK=ON
COUNTRY=001
SHELL=C:\COMMAND.COM /P /E:1024
LASTDRIVE=Z
```

Fig. 3. An example of a simple CONFIG.SYS file.

differs from the above we assume that you have sufficient familiarity with PC's and MS-DOS to apply our instructions to your system.

When MS-DOS is loaded after powering on your PC it looks for a series of files in a particular order. First MS-DOS looks for a CONFIG.SYS file and if it exists uses this file to configure the system for devices. Next MS-DOS loads the command processor, COMMAND.COM. The command processor looks for an AUTOEXEC.BAT file and if it exists the command processor executes the AUTOEXEC.BAT file. The ORACLE installation procedure requires that these three files exist and that they are located in the root directory.

The CONFIG.SYS File

The CONFIG.SYS file must be located in the root directory (C:\) of the MS-DOS system. The ORACLE installation process modifies this file so make sure it is not write protected (i.e. not read only mode). See 'Changing File Attributes' in this section for how to do this. After any changes to the CONFIG.SYS file MS-DOS must be re-started for the new commands to take effect. Figure 3 gives an example of a simple CONFIG.SYS FILE and the commands listed are discussed below.

DEVICE command

ANSI.SYS is a device driver that defines a standard set of methods for managing a screen display including how to display and erase characters, move the cursor and select colours. ORACLE requires that it be installed:

DEVICE = ANSI.SYS   ANSI.SYS must be in the root directory
DEVICE = C:\DOS\ANSI.SYS   ANSI.SYS is in the DOS directory

FILES command

FILES command sets the number of open files that MS-DOS can have access to at any one time.

FILES = 40   allows MS-DOS to have access
                           to up to 40 files at one time.

BUFFERS command

BUFFERS command sets the number of disk buffers allocated in memory when the system is started. MS-DOS uses these to handle reading from and writing to a disk.

BUFFERS = 20   sets up 20 disk buffers

BREAK command

BREAK command sets the [Ctrl][Break] check. [Ctrl][Break] can then be used in R mode to stop a program (the panic button).

COUNTRY command

COUNTRY identifies to MS-DOS which country's character set to use. ORACLE requires your system must be configured for American ASCII so make sure you have set up the country code for America:

COUNTRY = 001   COUNTRY.SYS must be in
                                          the root directory
COUNTRY = 001, ,C:\DOS\COUNTRY.SYS   COUNTRY.SYS is in the
                                          DOS directory

If an error is received when booting your system, or ORACLE will not load up after installation, then you may need to include the path in the COUNTRY command.

SHELL command

ORACLE makes extensive use of MS-DOS environment space. If you have already SET several commands and/or have a long PATH command in the AUTOEXEC.BAT file you may receive an 'out of environment space' error message from MS-DOS when you attempt to load ORACLE after installation. This means more environment space must be assigned. This can be achieved by inserting the following command into the CONFIG.SYS file:

SHELL = C:\COMMAND.COM /E:1024 /P

A word of warning: XTreePro is NOT compatible with MS-DOS Version 4 if a SHELL command is used. In this case when you boot your system you will receive a message flagging that something is wrong with your disk. In fact there is nothing wrong with your disk. XTreePro is fully compatible with earlier Versions of MS-DOS. XTreeGold is fully compatible with MS-DOS Version 4.

LASTDRIVE command

LASTDRIVE sets the maximum number of drives that may be accessed.

LASTDRIVE = Z   makes all the letters from a–z available as labels for
                       logical or physical drives

```
echo off
rem ===
rem ========= setup path =======================================
rem ===
PATH=C:\;C:\BATCH;C:\WS4;C:\XTPRO;C:\NORTON;C:\DOS;
cd\
rem ===
rem ========= setup prompt =====================================
rem ===
prompt $e[s$e[1;67H$e[31m$d$e[2;67H$t$h$h$h$e[u$e[36m$p $e[32m$g
rem ===
rem ========= ready to go ======================================
cls
```

Fig. 4.   An example of an AUTOEXEC.BAT file.

Changes to the CONFIG.SYS file can be made by changing to the root directory:

>CD \

and using an editor or word processor in non-document mode. Remember after making any changes to the CONFIG.SYS file to re-boot the system by simultaneously holding down the [Ctrl][Alt][Delete] keys for the changes to take effect.

The COMMAND.COM File

If the COMMAND.COM file is located in the root directory (C:\), then make sure that the root directory is one of the directories listed in the PATH command in the AUTOEXEC.BAT file. If, for some reason, the COMMAND.COM file is NOT in the root directory then add the following command to the AUTOEXEC. BAT file:

SET COMSPEC = C:\directoryname\COMMAND.COM

where directoryname is the directory that contains the COMMAND.COM file.

The AUTOEXEC.BAT File

The AUTOEXEC.BAT file must be located in the root directory (C:\). It contains commands that MS-DOS carries out when the system is started. Figure 4 shows an example of an AUTOEXEC.BAT file. It contains two commands that are useful if you keep MS-DOS and your programs in sub-directories: a PATH command and a PROMPT command. The PATH command tells MS-DOS the drive and directory path to follow when looking for external MS-DOS commands, application programs and programs from any directory on the disk. In the example the PROMPT command has been set up to display the current drive, the current directory and the time in the top right hand corner. It will not work unless ANSI.SYS is loaded.

A word of warning about the prompt command: check that the .EXE files of the various application packages have distinct names. If they are not you may find that issuing a command will have unexpected results if you are not invoking the software that you expect. For example ORACLE and Microsoft Fortran both use an EXP.EXE file. Entering EXP will invoke the software that is listed first in the PATH command. This can be overcome by re-issuing the PATH command from the MS-DOS command line. Experienced MS-DOS users should set up separate .BAT environment files in the BATCH directory for this purpose.

The ORACLE installation process amends the PATH command so make sure that the file is not write protected (i.e. that the read only mode is set). See 'Changing File Attributes' below for how to set the read only attribute. Whether the ORACLE RDBMS and its application tools are installed in P, R or B mode, the directories C:\ORACLE5\PBIN containing the P software code and C:\ORACLE5\BIN containing the R software code are added at the start of the path. If you have installed an application tool in B mode then, as the P mode directory appears first in the path, the tool will only ever execute in P mode unless you modify the PATH command so that the R mode directory is listed first. This can be done either by editing the AUTOEXEC.BAT file or re-issung the PATH command from the MS-DOS command line.

Changing File Attributes

To change the attribute of a file change to the directory containing the file, in this example the root directory (C:\):

> C >CD \

then

> C >ATTRIB filename.extn

If R, the read only attribute, is set then remove it by entering:

> C >ATTRIB −R filename.extn

Remember to re-set the R (read only) attribute on the files after the ORACLE installation process is complete:

> C >ATTRIB +R filename.extn

In Conclusion

If you are setting up your own MS-DOS system the CONFIG.SYS, COMMAND.COM and AUTOEXEC.BAT files must be in the root directory and we recommend that copies of the ANSI.SYS and COUNTRY.SYS files also be placed in the root directory (C: \). This makes installing the ORACLE RDBMS straight forward. If someone else has set up the system then some of these files may have been placed in directories other than the root directory. We have successfully installed ORACLE on several different types of machines (some ORACLE certified, many not certified) in different countries. We have encountered a variety of different system configurations and the advice given in this section is based on our experiences.

3. Installing the ORACLE RDBMS

The ORACLE installation procedure (OIP) has been updated considerably since the release of the earliest PC versions. The instructions given here are appropriate for ORACLE Version 5.1C and assume that you are installing the ORACLE RDBMS for the first time. If you are up-dating ORACLE or installing another version (other than 5.1C), refer to the documentation supplied with your version of the ORACLE RDBMS for the necessary instructions. The ORACLE installation procedure has two main parts:

1. System preparation to receive the ORACLE programs – running ORAINST from the floppy diskette drive A:
   The ORACLE directories are created and the CONFIG.SYS and the AUTO-EXEC.BAT files are modified. After making these changes you are prompted to reboot your system, if necessary.
2. ORACLE installation – running ORAINST from the hard disk. The ORACLE installation procedure prompts you to start installing the ORACLE programs beginning with the ORACLE Database followed by the ORACLE tools you wish to use.

3.1. ORACLE DIRECTORY STRUCTURE

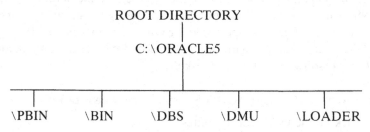

The ORACLE installation procedure creates a default directory named \ORACLE5 under the root directory (C:\). It also creates sub-directories in the C:\ORACLE5 directory to hold the files associated with different parts of ORACLE, as shown in the diagram above, namely:

\PBIN   holds the P mode executables for ORACLE RDBMS and the ORACLE Application Tools

\BIN	holds executable programs and batch files for ORACLE RDBMS and the ORACLE Application Tools
\DBS	holds the MS-DOS files that contain the ORACLE database, as well as several types of associated data files and parameter files
\DMU	holds files used by SQL*Menu
\LOADER	holds files used by SQL*Loader.

3.2. INSTALLING THE ORACLE RDBMS FOR MS-DOS

To install ORACLE RDBMS for MS-DOS, the ORACLE installation program called ORAINST is run.

ORAINST
- helps to prepare your system
- installs ORACLE products
- initialises ORACLE software
- can install update diskettes
- can list installed ORACLE products
- can remove ORACLE products

ORAINST is copied to the hard disk during installation. Additional ORACLE tools can be installed at a later time, by invoking ORAINST from the hard drive. ORAINST takes about 450K of hard disk space to run. If an error occurs during installation, press [Q] to quit the installation procedure (if necessary), remove any utilities or TSR programs currently resident in memory, then re-run ORAINST.

ORAINST will automatically modify the CONFIG.SYS and AUTOEXEC.BAT files, if necessary. So prior to running ORAINST check that the attributes of these files have not been set to read only. See Section III.2 (Preparing Your PC for the Installation of ORACLE RDBMS) that describes how to do this. You may choose to use an editor to modIfy these files yourself BEFORE running ORAINST. However, the simplest option, which is the one we describe here, is to elect to let ORAINST carry out the modifications for you during the installation process.

After your system has been prepared for ORACLE and rebooted, if necessary, re-run ORAINST from the hard disk. The second part of the installation copies the ORACLE software to the hard disk drive.

3.3. STEP BY STEP ORACLE RDBMS INSTALLATION INSTRUCTIONS

Step 1. Back up the ORACLE diskettes
- To protect your ORACLE software, make backup copies of all diskettes using the MS-DOS DISKCOPY command.
- See your MS-DOS reference manual for information on the DISKCOPY command.
- Do NOT use the COPY command to make the backups as it will not copy disk volume labels and OIP requires them.

- Label the copies carefully, then put the original diskettes in a safe place.
- Use the copies to perform the installation procedure.

Step 2. If you are re-installing ORACLE, back up your database to prevent loss of information. This can be done by using the EXP utility. Refer to Section III.6 (Backing Up and Restoring Your Database).

- for re-installation of the ORACLE RDBMS refer to the relevant documentation supplied with your version of ORACLE.

Step 3. Installing ORACLE for the first time

- Remember, the hard disk must have at least 8 Megabytes of free space to accommodate ORACLE RDBMS for MS-DOS and up to an additional 12 Megabytes for the application tools.
- Insert the diskette labelled 'Installation' in diskette drive A: and enter:

  C:\ >A:ORAINST

- You will be prompted for:
  - the drive from which the ORACLE products will be installed.
  - the drive and directory where the ORACLE products will be installed.
  - Press [Enter] to accept \ORACLE5 as the default directory in which you wish to install ORACLE.

Step 4. Modification of the CONFIG.SYS File

- ORAINST checks the CONFIG.SYS file that is read when you boot the system.
- If the successful operation of ORACLE products requires changes to the file, you are prompted to modify this file.
- To accept the automatic modifications press [Enter]. The CONFIG.SYS file will be similar to the one shown in Figure 3.

Step 5. Modification of the AUTOEXEC.BAT File

- ORAINST looks for an AUTOEXEC.BAT file on the boot drive specified in Step 3.
- If necessary, an AUTOEXEC.BAT file containing the required commands is automatically created, otherwise the necessary commands are appended to the existing AUTOEXEC.BAT file.
- To accept the automatic modifications press [Enter]. Figure 5 shows the modifications the installation procedure makes to the PATH command in the AUTO-EXEC.BAT file. See Section III.2 (Preparing Your PC for the Installation of ORACLE) for a discussion of the significance of the order of the ORACLE software sub-directories.

Step 6. List MACHTYPE's

- After the utilities are installed, SQLPME must know which hardware model you are using to proceed with installation.

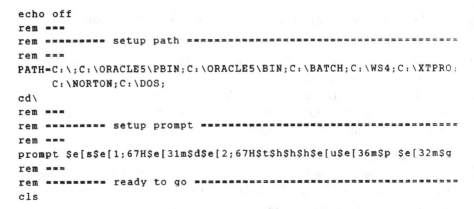

```
echo off
rem ===
rem ========== setup path ===================================================
rem ===
PATH=C:\;C:\ORACLE5\PBIN;C:\ORACLE5\BIN;C:\BATCH;C:\WS4;C:\XTPRO;
    C:\NORTON;C:\DOS;
cd\
rem ===
rem ========== setup prompt =================================================
rem ===
prompt $e[s$e[1;67H$e[31m$d$e[2;67H$t$h$h$h$e[u$e[36m$p $e[32m$g
rem ===
rem ========== ready to go ==================================================
cls
```

Fig. 5. The AUTOEXEC.BAT file of Figure 4 after installation of ORACLE.

- Press [C] to display a menu list of machines types, or press [Q] to Quit.
- If ORAINST has been previously run on your PC, ORAINST will detect the presence of SQLPME and there will be no prompt to carry out steps 6–7. Proceed directly to step 8 of the installation.

Step 7. Select the Hardware Model

- Use the arrow keys to highlight your hardware model and press [Enter] to accept.
- Hardware models that offer a sub-menu of model types are indicated by a "?". Press [Enter] to select the sub-menu and use the arrow keys to highlight the appropriate model, then press [Enter] to accept or [Q] to quit the sub-menu and return to the main menu.
- As you scroll through the list of hardware, note the number or letter in the "Machine Type" field. The machine type generally indicates the hardware model's CPU, so some of the models have the same machine type. For example, most 80286-based hardware models are MACHTYPE 0, most 80386-based hardware models are 2. However some are machine type 0 and some of the latest IBM PCs are MACHTYPE 6.
- If you select the wrong hardware model, the installation will be unsuccessful and you must reboot and re-install ORACLE RDBMS for MS-DOS. Now select the correct hardware model during the re-installation as described above.
- If you do not find your hardware model on the hardware menu list, try entering a machine type that most closely corresponds to the CPU of your hardware. Press [Esc] to move the cursor into the machine type field where the number or letter may be entered directly. ORACLE Database for MS-DOS can be successfully installed and run on non-certified models, or others not that are not listed, but ORACLE does not guarantee success.

Step 8. Rebooting

- If changes were made to the CONFIG.SYS and/or the AUTOEXEC.BAT files as a result of the installation procedure, you will be prompted to reboot.
- Remove the installation diskette from disk drive A: and simultaneously press the [Ctrl][Alt][Delete] keys to reboot the system.

Step 9. Run ORAINST from the Hard Disk
    Enter:
    C:\ >ORAINST

- You are prompted again to specify the drive from which products will be installed and the directory where they will be installed. Press Enter twice to accept the default settings.

Step 10. Choosing an Installation Task
- A menu of options is now displayed.
- Use the arrow keys to highlight "Install Product" in the list of installation tasks and press [Enter] to select this option.

Step 11. Select the products to install
- Insert the ORACLE Products Disk into floppy diskette drive A: when prompted and press [C] to continue.
- the products you need to install are:
  – Utilities
  – RDBMS
  – Tools Utilities
  – SQL*Plus
  – SQL*Forms
  – SQL*Menu
- Use the arrow keys to highlight these and press [Enter] to select each one from the list of products available.
    NOTE: – If the RDBMS is to be installed then the Utilities should also be installed.
          – If either or both of SQL*Forms and SQL*Menu are to be installed then Tools Utilities MUST also be installed.
- It is not necessary to install 'Client Manager' for use with the GPMDB. All its facilities and more have been incorporated into the custombuilt Menu Application, PMAG, supplied for use with the GPMDB.
- When your selection is complete press [C] to continue.
- If you wish to re-install a single product select only the one required.

Step 12. Questions Asked by the ORACLE Installation Program (OIP)
- OIP requires some information for most of the products you are to install. The program will prompt you for this
- The mode (R, P or B) in which you wish to install each product, except the RDBMS which can only be installed in P mode, must be specified.

- If the product you have selected is already installed, you may select "re-install" or "skip product".
- OIP verifies that there is enough hard disk space to install all the products you have selected. If there is not you will receive a warning message and should press [Q] to quit the OIP. Now either:
  - delete any unnecessary files from the hard disk and/or back up little used files to floppy diskette then delete them from the hard disk; pack the disk to create maximum contiguous file space; then restart ORAINST and repeat Steps 10–11;
  - or restart ORAINST and select fewer products to install. For example you may choose not to install SQL*Forms and/or SQL*Menu. Although the GPMDB can be accessed by SQL*Plus alone we would not recommend this as you will need to spend the time necessary to become very familiar with SQL. You will not be able to use the special packages we have developed using SQL*Forms and SQL*Menu that allow access to the GPMDB with minimum knowledge of SQL.
- Additional questions asked by OIP for the products you will load are as follows:

Utilities

RDBMS

OIP asks if you wish to install the default database, which is the initial database. You do, so press [Enter] to install it. If you have a non-certified machine and have specified the MACHTYPE incorrectly you may encounter difficulties in installing the RDBMS. The RDBMS ends with a message "initialising the database". This should only take about a minute to complete. If, after several minutes, you have not returned to the OIP and disk activity has halted, an unrecoverable error has occurred and you must reboot the system (simultaneously press the [Ctrl][Alt][Delete] keys). If you have an 80386 machine and chose MACHTYPE 2 you may have a system that is actually MACHTYPE 0. The simplest way to correct this problem is to enter the following commands at the MS-DOS prompt:

     C: \> CD C: \ORACLE5

     C: \ORACLE5> DIR

A file named CONFIG.ORA should be listed.

     C: \ORACLE5> TYPE CONFIG.ORA

Allows you to view the contents of the file. MACHTYPE = 2 should be one of the lines listed. Edit this line to change 2 to 0 and save the file. Return to the root directory (C:\) and re-run ORAINST repeating Steps 9–12. If this is not successful choose a new MACHTYPE and try again.

Tools Utilities

As you will be installing SQL*Forms and SQL*Menu, which are screen oriented, you must choose a default .CRT definition file. The available .CRT files are listed and described in your ORACLE documentation. Use the arrow keys to select the .CRT you require and press [Enter] to select it. We suggest BIOSBLUE as easiest on the eyes if it is compatible with your screen display type.

SQL*Plus

OIP asks if the help files are to be installed. As you have already installed the RDBMS, the help files were loaded in this step. Answer "No".

SQL*Forms

OIP asks whether to install the SQL*Forms software for designing Forms applications. If you plan to design your own Forms answer "Yes". However if you intend to run only the custombuilt Form PALEOMAG that we supply with GPMDB then answer "No".

OIP asks if you plan to use "user exits". Answer "No". (We have not made use of this facility in the GPMDB and the software will take up extra hard disk space). OIP asks if you wish to install the database tables required to run SQL*Forms. Answer "Yes".

SQL*Menu

OIP asks if you wish to install the tables required by SQL*Menu. Answer "Yes". OIP gives you the option of installing a sample Personnel Application. Answer "No".

Step 13. Completion of the ORACLE Installation Process

- After all the necessary files have been copied to the hard disk, you are automatically returned to the start of the installation program.
- Press [Q] to quit the OIP and exit to MS-DOS.

4. System Preparation and Loading the Global Paleomagnetic Database

4.1. SYSTEM PREPARATION

Your system must be prepared before installing the GPMDB. Load ORACLE if not already loaded:

```
C:\ >ORACLE
```

Re-Initialising the Database

First re-initialise the database:

C:\ >IOR I

You will be requested to confirm your intention to re-initialise as all data in the database will be lost when the initialisation takes place.

"IOR connecting to ORACLE V5.0

Initialise of entire database requested. All data will be lost.

– Confirm (YES/NO) ?"

Answer YES in upper case.
The initialisation will take some time.

When the database has been initialised the following message will appear:

"DATABASE INITIALISED"

Creating a New Database File

Change to the ORACLE home directory:

C:\ >CD\ORACLE5\DBS

Among other files this directory contains the MS-DOS files automatically created by ORAINST into which the database will be loaded. ORACLE 5.1C is supplied with 2 database files, DBS.ORA and DBS2.ORA that total approximately 3 Megabytes. This is not large enough to contain the GPMDB and allow querying of it. A new database file must be created. The following steps have been found by us to cause the least confusion and are least error prone. They differ from the standard ORACLE procedure but make importing the GPMDB and importing a new version of the database follow the same steps.

If you are importing a new version of the GPMDB make sure you back up any tables you and your users have created before re-initialising. Views, indexes, grants and synonyms created by you or your users must be recreated after re-initialising and importing the new database

ALL THE FOLLOWING OPERATIONS MUST BE CARRIED OUT IN DIRECTORY C:\ORACLE\DBS.

To delete DBS2.0RA then create and add a new DBS2.0RA file to directory C:\ORACLE5\DBS enter the following commands:

C:\ORACLE5\DBS >DEL DBS2.ORA

C:\ORACLE5\DBS> EXPAND SYSTEM/password DBS2.ORA 12288

Take great care to duplicate this statement exactly. This gives a database of about 8 Megabytes in total, and should be adequate until the GPMDB is updated in 1992.

If you make a mistake entering the EXPAND command you will need to delete the DBS2.ORA file, if it exists, re-initialise the database, and re-issue the EXPAND command correctly.

Creating a New Before Image File

As part of the initialisation process ORACLE creates a Before Image (BI) file of approximately 1 Megabyte. This file is used by ORACLE to protect the integrity of the database during transactions. If it is too small, an error will result when queries that retrieve a large number of rows are run. As the GPMDB is considerably larger than the default database supplied by ORACLE, a new larger BI file must be created. The rule of thumb is that the BI file should be $\frac{1}{3}$ to $\frac{1}{2}$ the size of the total database files. There is a total of about 8 Megabytes of database files so we will create a BI file of 4 Megabytes.

First back up the current BI file onto a floppy diskette:

ALL THE FOLLOWING OPERATIONS MUST BE CARRIED OUT IN THE DIRECTORY C:\ORACLE5\DBS.

> C:\ >CD C:\ORACLE5\DBS
> C:\ORACLE5\DBS >XCOPY BI.ORA   A:BI.ORA

With ORACLE loaded:

1. C:\ORACLE5\DBS >IOR S   shut down the database

If the following message appears: "No active transactions found" then skip Steps 2 and 3 and go to Step 4.
2. C:\ORACLE5\DBS >IOR W DBA   re-start the database
3. C:\ORACLE5\DBS >IOR S   then shut it down again to clear the
                                                           BI file
4. Rename the existing BI file:

> C:\ORACLE5\DBS >REN BI.ORA   BI.OLD

5. Create a new BI file with the same name as the original file:

> C:\ORACLE5\DBS >CCF BI.ORA   8192

6. Now re-start ORACLE with the new BI file:

> C:\ORACLE\DBS >IOR W

7. If ORACLE re-starts successfully then your system is now ready to load the GPMDB. This is done via the ORACLE import utility.
8. If not, check that file BI.ORA exists and is in the C:\ORACLE5\DBS directory. If there is no BI.ORA file then check that there is sufficient space on the disk and repeat Steps 5 and 6.
9. Delete BI.OLD to release the hard disk space.

4.2. INSTALLING THE GLOBAL PALEOMAGNETIC DATABASE

You should have been supplied with 3 or more diskettes:
1. 2 or more high density diskettes that contain the GPMDB
   - Diskette 1, 2 etc.
2. 1 low density diskette that contains MS-DOS files
   - Diskette A.

If you have installed an early version of ORACLE (5.1, 5.1A or early 5.1B) you should contact your ORACLE dealer and arrange to have an update of the IMP and EXP utilities supplied. If you have installed ORACLE 5.1C or later versions of 5.1B then the GPMDB can be imported directly from the floppy diskettes using the procedure described below. Place diskette 1 (high density) in drive A: and with ORACLE loaded enter the command:

C:\PALEOMAG >IMP SYSTEM/password

This is an interactive utility that asks a series of questions, each terminated by a > sign. Press [Enter] to accept the default. Answer each question as shown in Table V.

If you receive system error messages during the IMPORT check that you have configured your system for American ASCII (refer to Section III.2 – Preparing Your PC for the Installation of the ORACLE RDBMS). Table V shows the table, view, and grant list generated by the IMPORT procedure. The IMPORT will take up to an hour or even more depending on your PC hardware model.

Notes on the IMPORT procedure:

1. If you receive a message that you are not the owner of the data, DO NOT PANIC. You are indeed not the owner of the data as you did not create it. This does not prevent you from successfully importing the data onto your system and querying it. You can ignore this message and the import utility will continue with its questions.
2. If the database import does not go to completion and gives an error message, it is likely that you did not EXPAND the ORACLE database correctly. Refer back to Section III.4.1 (Creating a New Database File).

The GPMDB is now ready for use.
We have built in protection for the GPMDB by restricting authorised users'

TABLE V
Complete set of messages received during IMPort of the Global Paleomagnetic Database IMPORT
files (Numbers of rows refer to the January 1991 version of the GPMDB)

Import file: EXPDAT.DMP > A:GPMDB.DMP   [Enter]
Enter insert buffer size (default 10240, minimum 4096) > [Enter]
Export created by ORACLE version EXPORT: V05.01.22
List contents of import file (Y/N): N > N   [Enter]
Ignore create errors due to object existence (Y/N): N > Y   [Enter]
Import Grants (Y/N): Y > Y   [Enter]
Import the rows (Y/N): Y > Y   [Enter]

Import of entire import file requested (Y/N): Y > Y   [Enter]
.Importing user SYSTEM
..Importing table "AUDIT_ACTIONS"   49 Rows imported
..Importing table "DTAB"   42 Rows imported
..Importing table "DUAL"   1 Rows imported
..Importing table "MENU_APPLICATION"   2 Rows imported
..Importing table "MENU_COMMAND_TYPE"   6 Rows imported
..Importing table "MENU_HELP"   250 Rows imported
..Importing table "MENU_INFO"   12 Rows imported
..Importing table "MENU_MESSAGE"   172 Rows imported
..Importing table "MENU_OPTION"   80 Rows imported
..Importing table "MENU_PARAM"   14 Rows imported
..Importing table "MENU_PARAM_XRFF"   0 Rows imported
..Importing table "MENU_USER"   4 Rows imported
..Importing table "MENU_WORK_CLASS"   5 Rows imported
.Importing user DATABASE
..Importing table "ALTRESULT"   847 Rows imported
..Importing table "AUTHORS"   1863 Rows imported
..Importing table "CROSSREF"   3095 Rows imported
..Importing table "FIELDTESTS"   819 Rows imported
..Importing table "INFORMATION"   75 Rows imported
..Importing table "JOURNAL"   35 Rows imported
..Importing table "KEYS"   20 Rows imported
..Importing table "PMAGRESULT"   3786 Rows imported
..Importing table "REFERENCE"
End of file: insert next diskette and enter the name
of the next file (default A:GPMDB.001), or . to abort: [Enter]
    1863 Rows imported
..Importing table "REMARKS"   1863 Rows imported
..Importing table "ROCKUNIT"   3355 Rows imported
..Importing table "TIMESCALE"   58 Rows imported
.Importing user WORKSHOP
.Importing user SYSTEM
.Importing user DATABASE
.Importing user SYSTEM
.Importing user DATABASE
.Importing user SYS

privileges to SELECT only on any of the database tables. Querying the database may be done either by using a SQL*Plus SELECT command, via the customised Form supplied with the database (on diskette A), or via the customised Menu Application. Updating data (changing existing entries), deleting data, or committing of new entries to the database are not permitted. If omissions or errata are found please notify us in writing and we will arrange for these corrections to be made available with the next update of the database.

4.3. LOADING THE CUSTOMBUILT FILES

This document describes how to load the customised batch .BAT files, SQL command .SQL files, the LOGIN.SQL file and customised query form, PALEO-MAG supplied with the GPMDB on diskette A (low density).

If you have purchased and loaded a disk management package such as XTree-Pro, then the following steps can be very easily accomplished using it. Refer to the documentation supplied with the package. The instructions given here are for MS-DOS.

Create two new sub-directories in the root directory (C:\):

```
> CD \              (if necessary change to the root directory)
C:\ >MD BATCH       (if you do not already have one)
C:\ >MD PALEOMAG    (this will be your ORACLE work directory)
```

Installing the .BAT Files

Place diskette A in drive A: and enter the following commands:

```
C: \ >CD \BATCH
C:\BATCH >XCOPY A:*.BAT
```

This MS-DOS command copies all files on the floppy diskette with the extension .BAT to your current dIrectory the BATCH directory.

Make sure that C:\BATCH; has been added early in the PATH command of the AUTOEXEC.BAT file in the root directory. To do this enter the following commands:

```
C:\ BATCH >CD C:\
C:\ >TYPE AUTOEXEC.BAT
```

These commands change to the root directory, and list the contents of the AUTO-EXEC.BAT file. If the PATH statement does not include C:\BATCH; the file will need editing. The attribute of the file may need to be changed from read only. This can be done by entering the following command:

```
C:\ >ATTRIB −R AUTOEXEC.BAT
```

Now edit the file to include C:\BATCH; early in the PATH statement and exit the editor. Reset the read only attribute by entering the following command:

> C:\ >ATTRIB +R AUTOEXEC.BAT

If you changed the PATH command then reboot (simultaneously press the [Ctrl][Alt][Delete] keys) for the change to take effect.

Except for the AUTOEXEC.BAT file, which must be in the root directory, the BATCH directory should be used to store any other .BAT files that you create.

From any directory that is a sub-directory of the root directory, enter the following command:

> >GORACLE

then ORACLE will be STARTED and the directory will be automatically changed to C:\PALEOMAG, your GPMDB work directory.

Installing Custombuilt .SQL, .FRM and .DMM Files

To install the customised files supplied with the GPMDB enter the following commands:

> C >CD \PALEOMAG
> C: \PALEOMAG >XCOPY A:*.SQL
> C: \PALEOMAG >XCOPY A:*.FRM
> C: \PALEOMAG >XCOPY A:*.DMM

These commands will copy all the customised SQL command files, the customised Form and the customised Menu from the diskette to your hard disk into the GPMDB work directory, C:\PALEOMAG. The * character acts as a wild card and will select any file from the diskette which has an extension matching that specified. Appendix 2 lists the code contained in the customised .SQL command files that are supplied with the GPMDB.

The print version of the .SQL command files (prefixed by a 'P') assume that the printer port is LPT1. If your printer is connected to another LPT port or a COM port, the LPT1 print output can be re-directed to that port by adding the appropriate commands to the LOGIN.SQL file as described in the next section.

The LOGIN.SQL File

The LOGIN.SQL file is consulted by SQL*Plus each time a user logs into SQL and sets parameters for use within SQL. ORACLE supplies two versions of this file:

LOGIN.SQL   . gives prompt SQL>
               . sets up pagewidth and pagesize

. headings are printed at the start of every page

LOGIN.NEW . gives a colour screen with prompt ORACLE> in red

. text is in grey on a black background

. sets up pagewidth and pagesize

. sets an automatic pause at the end of each page of output and prints "More" in blue at the bottom right hand corner of the page. Press [Return] to see the next page. This may not work on all systems.

Choose which version you prefer then:

    C:\PALEOMAG >CD \ORACLE5

    C:\ORACLE5 >XCOPY LOGIN.SQL

        C:\PALEOMAG\LOGIN.SQL

or

    C:\ORACLE5 >XCOPY LOGIN.NEW

        C:\PALEOMAG\LOGIN.SQL

SQL*Plus has a command EDIT which allows editing from within SQL, however the default is the MS-DOS editor EDLIN. We find this a very unfriendly beast and prefer to use WORDSTAR (you may wish to use another editor or word processor). If you wish to change this default, change to the work directory:

    C:\ORACLE5 >CD \PALEOMAG

then edit LOGIN.SQL by inserting after the SET commands:

    DEFINE _EDITOR ='xx'

where 'xx' is the command that you normally use to log into your favourite editor or word processor. If you use a word processor make sure the output text is a normal ASCII file by using the mode that does not automatically include formatting and printing commands in the output file. In WORDSTAR this is non-document mode and 'xx' is 'ws'.

Other SQL*Plus commands can be inserted to customise SQL to your needs and these will take effect at each login. For example, if your printer is connected to LPT2, to re-direct output from LPT1 insert the following command into the LOGIN.SQL file:

    MODE LPT2:

If your printer is connected to the COM1 port, output from LPT1 can be re-directed by inserting the following commands into the LOGIN.SQL file:

    MODE COM1:

    MODE LPT1: = COM1:

```
REM Example LOGIN.SQL invoking color via the use of the ANSI.SYS
REM device driver
set numwidth 7
set linesize 79
set pagesize 25
define_editor = 'ws'
set sqlprompt ';31mORACLE>m'
set pause '5;72H4mHore;5m...;32m' pause on
```

Fig. 6.   The LOGIN.SQL file used by the authors. (Note: This form of prompt and pause may not work on all display types).

Figure 6 shows the LOGIN.SQL that we use.

Checking the Custombuilt Form for your Display

If you followed our recommendations when SQL*Forms was loaded the .CRT file selected was BIOSBLUE. Those function key definitions for SQL*Forms and SQL*Menu are used in this document. To check that the customised Form will work correctly on your particular CRT enter the following command:

C: \PALEOMAG >RUNFORM PALEOMAG userID/password

You should see two boxes one underneath the other labelled AUTHORS and REFERENCE that contain labelled highlighted fields as shown in Figure 12 (Section IV). If this is the case, exit RUNFORM by pressing [Esc] or [Shift][F10] keys, skip the next section and go to 'Access to the Custombuilt Files'.

Re-creating the Custombuilt Form for your Display

The customised Form has been compatible with all the varied PC systems we have used in different countries. Therefore, we do not anticipate that you will need this section!

If you did not load the SQL*Forms Designer when installing ORACLE, then re-run ORAINST and install SQL*Forms in Real or Protected mode depending on the amount of *EXTENDED* memory available. There are two potential sources of Form incompatibility. Firstly, while the database and .INP form file formats are the same on every type of computer system that supports SQL*Forms, the .FRM file format may not be. The following procedure will produce .FRM files compatible with your system.

Exit RUNFORM by pressing the [Esc] key.

With diskette A in drive A: enter the following commands:

C: \PALEOMAG >XCOPY A:*.INP
C: \PALEOMAG >IAG formname −TO

where formnames are: PALEOMAG
JOURNAL

TIME
INFO
KEYS

If no errors are reported, when the MS-DOS prompt appears enter:

C:\PALEOMAG >RUNFORM PALEOMAG userID/password

and look for the characteristics mentioned above.

If your screen still does not conform to this description then exit RUNFORM by pressing the [Esc] key. The second potential source of incompatibility is that characteristics of your screen display may be different to those of the screen on which the Form was developed. The .INP and .FRM Form files store lines and boxes as a sequence of text characters. These are device dependent. All of the Form supplied with the GPMDB uses boxes. If the boxes are missing and/or there are strange characters on your screen you must create a new version of each database file. Starting with the current .INP Form file, convert and load this file into the database then recreate .INP and .FRM Form files that are compatible with your screen by entering:

C:\PALEOMAG >IAC −IS Formname Formname SYSTEM/password
C:\PALEOMAG >IAC Formname Formname SYSTEM/password
C:\PALEOMAG >IAG Formname −TO

Repeat these commands for each of the five Formnames above. Note that Formname appears twice in the first two commands. These refer to different versions of the Form files (Database, .INP, .FRM). We have given the three versions of the Form files the same name. ORACLE knows from the position of the name in the command which version of the Form file is to be used.

C:\PALEOMAG >RUNFORM PALEOMAG userID/password

should now look like Figure 12. To remove each of form files from the database and release space they occupy in the ORACLE database files enter:

C:\PALEOMAG >IAC −D Formname Formname SYSTEM/password

This does not affect the .INP and .FRM files as these are stored as MS-DOS files outside the database. The steps used to create the compatible RUNFORM version of each Form need only be carried out once, at installation. If you do not plan to design your own Forms Applications you may wish to remove the SQL*Forms Designer (refer to Section II.2.1 − Dropping the SQL*Forms Extended Data Dictionary).

Access to the Custombuilt Files

Any authorised ORACLE user has access to these customised files provided they are working in the ORACLE work directory C:\PALEOMAG, or specify the full path name to this directory as well as the Form name or .SQL command file name.

5. Database Administrative Tasks

On a PC ongoing DBA tasks are:

- Setting up userID's and passwords for new users
- Dropping old users
- Checking of available database storage
- Expanding the database
- Creating a new BI file
- EXPorting the GPMDB to back up tables, views and indexes created by users of the GPMDB
- IMPorting updates of the GPMDB

5.1. STARTING THE ORACLE RDBMS

ORACLE supplies a batch file, ORACLE, that can be used to start ORACLE from any subdirectory of the root directory (C:\). To start ORACLE, at the MS-DOS prompt enter the command:

C:\ >ORACLE

ORACLE • invokes SQLPME
- loads the ORACLE kernel into extended memory
- IOR W, warm starts ORACLE; it
  - opens the database
  - opens the Before Image (BI) file
  - starts ORACLE background processes
- displays System Global Area (SGI) information
- displays IOR: instance 1 or 2 recovered.
  IOR: ORACLE warm started.

If you have a non-certified PC and on entering ORACLE, the above does not run to completion, or an error message is displayed, the most likely cause is that the MACHTYPE has been incorrectly specified. First try re-specifying the MACHTYPE at the MS-DOS prompt by entering:

C:\ >MACHTYPE

and following the instructions given in step 7 of Section III.3 (Installing the

ORACLE RDBMS) to choose a different hardware model. Then try to start ORACLE by entering:

> C:\ >ORACLE

If you are still unsuccessful in starting ORACLE then either you still have the wrong MACHTYPE or the initial incorrect specification of MACHTYPE caused some ORACLE software to be installed incorrectly. This can only be corrected by removing all the ORACLE software files, deleting the sub-directories and the ORACLES directory, and repeating the installation process again in its entirety.

A further memory count check should be made as ORACLE is loading into extended memory from the hard disk where it has been installed. The amount of extended memory that is counted and displayed as being available should be the same as the amount available for use by ORACLE utilities. If only 1 Megabyte of extended memory is counted then the utilities and tools should have been loaded in R mode. If they were loaded in P mode, error messages will result when you try to log into SQL*Plus. If you have chosen the wrong mode then shut down ORACLE and remove it from extended memory by entering the following commands:

> C:\ >IOR S
> C:\ >REMORA ALL
> C:\ >REMPME

Now re-run ORAINST and install the ORACLE utilities and tools in R mode.

We have supplied two customised versions of the ORACLE batch file for use with the GPMDB. Instructions on how and where to load these files are given in Section III.5.3 (Loading the Custombuilt Files).

GORACLE – carries out the same steps as the ORACLE batch file.
– changes from the the directory you are currently in to the GPMDB work directory, C:\PALEOMAG.
LOADORA – carries out the same steps as the GORACLE batch file.
– logs the user into SQL*Menu and prompts for the userID and password.

On startup ORACLE also consults two files, C:\ORACLE5\CONFIG.ORA and C:\ORACLE5\DBS\INIT.ORA, that control MS-DOS system resources and set the general environmental characteristics respectively.

If you attempt to log into an ORACLE utility or tool and ORACLE has not been started then you will receive the message "PME services are not available".

5.2. STOPPING THE ORACLE RDBMS

If ORACLE has not been shut down and you, or another user, attempt to re-load ORACLE, an error message will be received:

C:\ >"Error: SQLPME is already loaded"

This does not make it clear that ORACLE is already loaded. To avoid this situation it is good practice to shut ORACLE down when you have finished a database session. In any case ORACLE should *ALWAYS* be shut down before switching off your PC.

At the MS-DOS PROMPT, enter the command;

C:\ >IOR S   • closes the database file
             • closes the BI file
             • stops the ORACLE background processes
             • Displays IOR: No active transactions found
                       IOR: ORACLE shutdown complete
             • ORACLE still remains in the extended memory

After issuing the IOR S command you can shut down your system at any time. It is not necessary to remove ORACLE from extended memory before shutting down your PC.

Once ORACLE has been started (IOR W) it remains in memory even when an individual program ends and the MS-DOS prompt is displayed – i.e. ORACLE is a terminate and stay resident (TSR) program. The majority of ORACLE software runs in extended memory. You may wish to free this memory for use by other programs. We have supplied a batch file called SHUTORA that is designed to shut ORACLE down and remove it from extended memory.

C:\ >SHUTORA   • IOR S
                 – shuts ORACLE down as described above
               • REMORA ALL
                 – removes the currently loaded driver
                 – removes the ORACLE RDBMS from RAM
                 – frees all extended memory used by RDBMS
               • REMPME
                 – frees all extended memory used by SQLPME, the ORA-
                   CLE kernel and any other other ORACLE product cur-
                   rently loaded.

5.3. CHANGING THE SYSTEM DEFAULT PASSWORDS

ORACLE RDBMS is initially installed with two userID's that have DBA privileges. They have default passwords that, to protect your system should be changed immediately after installation.

SYS   • owns the data dictionary tables that are updated automatically by
            ORACLE. You *MUST NOT* alter these tables in any way. *NEVER*
            log on as user SYS except once to change the default password.

SYSTEM ● owns the data dictionary views based on the tables. Altering these views is possible but *NOT* recommended. A user with DBA privileges can modify certain tables that are critical to the operation of ORACLE. Log on as SYSTEM only to carry out system administration functions.

To change the default passwords start ORACLE and log into SQL*Plus by entering the following commands at the MS-DOS prompt:

```
C:\ >GORACLE
C:\PALEOMAG >SQLPLUS SYS/CHANGE_ON_INSTALL
SQL> GRANT CONNECT TO SYS IDENTIFIED BY
     newpasswordsys;
SQL> CONNECT SYSTEM/MANAGER
SQL> GRANT CONNECT TO SYSTEM IDENTIFIED BY
     newpasswordsystem;
SQL> EXIT
C:\PALEOMAG >
```

Be sure to make a record of the new SYS and SYSTEM passwords and keep them in a safe place so the unauthorised users do not gain access to them. BEWARE, ORACLE encodes each password in such a way that it is impossible to decode them. Do not trust passwords to memory, always record them.

5.4. SETTING UP USERIDS AND PASSWORDS

ORACLE security requires that every user log on with their own userID and password provided by the DBA. A userID and password is required to log into the various ORACLE tools such as SQL*Plus, SQL*Forms and SQL*Menu. A unique userID and password allows a user to create his own tables and views of the GPMDB to which other users have no access unless it is granted by their owner (creator). UserID's and passwords are set using SQL*Plus's GRANT command. This has general syntax as follows:

GRANT privileges TO userID IDENTIFIED BY password;

The privileges that may be granted are:

CONNECT   allows the user to connect to the database
RESOURCE   allows the user to create tables, views and indexes
DBA   allows the user to perform DBA functions. This privilege is not
           to be granted to the ordinary user

NOTE: There is a semi-colon at the end of the GRANT command. A semi-colon must terminate each SQL command. This instructs SQL to run the command.

For example, to grant a userID to Mike, start ORACLE and log into SQL* Plus as the DBA SYSTEM by entering the following commands at the MS-DOS prompt:

```
C:\>GORACLE                (if necessary load ORACLE)
C:\PALEOMAG >SQLPLUS SYSTEM/password
SQL> GRANT CONNECT, RESOURCE TO MIKE IDENTIFIED
     BY MWM;
SQL> EXIT
C:\PALEOMAG >
```

If Mike decides to change his password from MWM to GONDWANA he may do so at any time by logging into SQL*Plus and entering:

```
C:\PALEOMAG >SQLPLUS MIKE/MWM
SQL> GRANT CONNECT TO MIKE IDENTIFIED BY
     GONDWANA;
SQL> EXIT
C:\PALEOMAG >
```

Passwords should be changed regularly to maintain security.

ORACLE automatically maintains tables and views that comprise the Data Dictionary. These tables may be queried to find current information about the database. For example, to obtain a list of current users you may query the Data Dictionary Table SYSUSERLIST from SQL*Plus using your userID and password by entering:

```
C:\PALEOMAG >SQLPLUS UserID/password
SQL> SELECT *
   2 FROM SYSUSERLIST;
SQL> EXIT
C:\PALEOMAG >
```

When the GPMDB is installed the users shown in Figure 7 are IMPorted with the database. Note that a general user WORKSHOP with password PMAG has been provided. This will assist with the initial operation of SQL*Menu (see discussion below). In addition the user DATABASE has been IMPorted. DATABASE is the creator and owner of all the GPMDB tables, views and indexes. Although you will never be able to logon as DATABASE without the correct password *DO NOT* revoke DATABASE as an authorised ORACLE user. If you were to do this you would lose all access to the GPMDB itself.

The DBA may revoke database privileges from a user. For example:

```
C:\PALEOMAG >SQLPLUS SYSTEM/password
SQL> REVOKE RESOURCE FROM MIKE;
```

```
USERID USERNAME                             TIMESTAMP C D R
------ ------------------------------------ --------- - - -
   5                                        01-JAN-91
   0 SYS                                    01-JAN-91 Y Y Y
   1 PUBLIC                                 01-JAN-91
   2 SYSTEM                                 01-JAN-91 Y Y Y
   3 DATABASE                               11-JAN-91 Y   Y
   4 WORKSHOP                               13-JAN-91 Y   Y

6 records selected.
```

Fig. 7.   List of users imported with the GPMDB, as displayed by SYSUSERLIST. C (Connect), D (DBA), R (Resource) indicate the privileges granted with symbol Y.

reduces Mike's privileges to connecting to the database. He can no longer create tables, views and indexes. He may be dropped altogether as a user by revoking his connect privilege:

SQL> REVOKE CONNECT FROM MIKE;

Both privileges could have been removed at the same time:

SQL> REVOKE CONNECT, RESOURCE FROM MIKE;

Notice that the DBA does not need to know Mike's password to REVOKE privileges.

   When you have finished your database session don't forget to shut down ORA-CLE. You must exit from any ORACLE tool you were using before attempting to shut down ORACLE. At the MS-DOS prompt enter:

C:\PALEOMAG >SHUTORA

   To log into SQL*Menu every user must also be granted access to SQL*Menu and in addition must be given a Work_Class assignment for the Menu Application to be used. For this purpose the GPMDB provides a general user with userID WORKSHOP and password PMAG. WORKSHOP is automatically imported as an ORACLE user when the GPMDB is first IMPorted. At the same time WORK-SHOP is installed as an authorised SQL*Menu user and given a Work-Class assignment

   NOTE: Only user WORKSHOP/PMAG will initially have access to and use of the GPMDB Menu PMAG. SYSTEM is the ONLY user who may grant access and Work_Class assignments to the Menu PMAG as described below.

5.5. SETTING UP USER ACCESS TO SQL*MENU

*Logging into SQL*Menu*

A custombuilt menu called PMAG has been developed for easy use of all the packaged programs created for the GPMDB. Once you have logged into

SQL*Menu all ORACLE or MS-DOS commands can be executed from the Menu. To start up ORACLE at the beginning of a session enter the following command at the MS-DOS prompt:

C:\ >LOADORA

This will load ORACLE into extended memory and put the user into SQL*Menu where you will be prompted for your UserID and Password. The Menu as installed with the database automatically comes with a general user called WORKSHOP with password PMAG.

Every user, in addition to being granted access to ORACLE, must also be granted access to SQL*Menu and be given a Work_Class Assignment for use of the Application Menu PMAG as described below.

*Granting New Users Access to SQL*Menu and Application PMAG*

Each user can only be granted access to SQL*Menu by the DBA through the user SYSTEM/password. If already in SQL*Menu you should Exit using the Menu and reload SQL*Menu using SYSTEM/password. If ORACLE is already loaded then at the MS-DOS prompt enter:

C:\PALEOMAG >SQLMENU

You will then be prompted for your userID/password (SYSTEM/password). The Application Menu will be displayed as shown in Figure 8(a).

Application Menu

Select Option 1 – SQL*Menu Development of Dynamic Menus.
      The SQL*Menu Dynamic Utility Main Menu will now be displayed as illustrated in Figure 8(b).

*SQL*Menu Main Menu*

Select Option 3 – SQL*Menu System Maintenance
      The DBA Menu is displayed as in Figure 8(c).

Database Administrator Menu

Select Option 4 – Granting of a New User of SQL*Menu
      The option parameter form shown in Figure 8(d) will be displayed. Now enter the userID to be granted access to SQL*Menu. Press [Enter], leave the Grant Option window blank and press [Enter] again. SQL*Menu will first check if you are a registered ORACLE user and then return you to the DBA Menu.

(a)
```
                    A P P L I C A T I O N    M E N U

                SELECT AN APPLICATION FROM THE LIST BELOW


            -->   1   SQL*Menu Development of Dynamic Menus
                  2   GLOBAL PALEOMAGNETIC DATABASE
                  3   Exit




                      Make your choice:  1
```
_____

```
v    Sun Feb 03 10:04:57 1991                 Replace       ($apl$ )
```

(b)
```
        S Q L * M e n u    D y n a m i c    M e n u    U t i l i t y

                        SQL*Menu Main menu


              1   Create and Maintain SQL*Menu Applications
              2   Generation of SQL*Menu Documentation
          --> 3   SQL*Menu System Maintenance
              4   Previous Menu




                      Make your choice:  3
```
_____

```
v ^ Sun Feb 03 10:05:51 1991      BGH OSC DBG      Replace      DMU (DMU )
```
                             Fig. 8(a, b).

Database Administrator Menu

Select Option 6 – Previous Menu
> This will return you to the Main Menu as shown in Figure 9(a).

You must now grant the user a Work_Class assignment in the Application Menu
PMAG.

(c)
```
S Q L * M e n u    D y n a m i c    M e n u    U t i l i t y

                 Database Administrator Menu


              1  Creation of a New SQL*Menu System Library
              2  Maintenance of Message Database
              3  Creation of a New Language Definition
        -->   4  Granting of a New User of SQL*Menu
              5  Creating a Skeleton for a New Language
              6  Previous Menu




                 Make your choice:  4

     Options on this menu are for tailoring general appearance of SQL*Menu
```
v ^ Sun Feb 03 10:06:36 1991   BGM OSC DBG   Replace   DHU (DMUDBA)

(d)
```
            O P T I O N    P A R A M E T E R    F O R M


                    Grantee   WORKSHOP

                 Grant option
```


Sun Feb 03 10:07:05 1991   Replace   page 1 : 1

Fig. 8.   Method of granting new users access to SQL*Menu. (a) Application Menu (SYSTEM only). Select Option 1. (b) SQL*Menu Main Menu. Select Option 3. (c) Database Administrator Menu. Select Option 4. (d) Option Parameter Form. Enter grantee userID and leave Grant option blank.

SQL*Menu Main Menu

Select Option 1 – Create and Maintain SQL*Menu Applications

The Menu Information Maintenance will now be displayed as shown in Figure 9(b).

(a)

```
S Q L * M e n u    D y n a m i c    M e n u    U t i l i t y

                 SQL*Menu Main menu


         --> 1   Create and Maintain SQL*Menu Applications
             2   Generation of SQL*Menu Documentation
             3   SQL*Menu System Maintenance
             4   Previous Menu




                     Make your choice:  1
```
_____

```
v    Sun Feb 03 10:08:01 1991     BGM OSC DBG        Replace      DMU (DMU )
```

(b)

```
         S Q L * M e n u    D y n a m i c    M e n u    U t i l i t y

                       Menu Information Maintenance


                  1   Update Application Information
           --> 2   Update Work-class and User Information
                  3   Update Menu Information
                  4   Update Substitution Parameter Information
                  5   Generate One Menu
                  6   Generate All Menus for One Application
                  7   Manage Libraries and Applications
                  8   Previous Menu




                       Make your choice:  2

       NOTE: The meaning of function keys can change when running SQL*Forms
```
_____

```
v  ^ Sun Feb 03 10:08:41 1991     BGM OSC DBG        Replace      DMU (DMUMNU )
```

Fig. 9(a, b).

Menu Information Maintenance

Select Option 2 – Update Work-Class and User Information

This will put you into the Work-Class/User Information Form
as shown in Figure 9(c).

(c)

```
          ┌──────────────────────────────────────────────┐
          │    SQL*MENU - WORK CLASS/USER INFORMATION     │   03-FEB-91
          │                                              │
          │      Application_name: PMAG                   │
    ┌─────┴──────────────────────────────────────────────┴─────────┐
    │ Work_class : 5                                                │
    │ Description: General Application User                         │
    │                                                               │
    │                                                               │
    └──────┬──────────────────────────────────────────────┬────────┘
           │                                              │
           │   Work_class   User_name        DBG OSC BGM  │
           │   5            WORKSHOP          N   Y   Y    │
           │                                              │
           │                                              │
           │                                              │
           │                                              │
           │                                              │
           └──────────────────────────────────────────────┘
```

v Char Mode: Replace   Page 1   Count: 1

(d)

        S Q L * M e n u   D y n a m i c   M e n u   U t i l i t y

                    Menu Information Maintenance

             1 Update Application Information
             2 Update Work-class and User Information
             3 Update Menu Information
             4 Update Substitution Parameter Information
             5 Generate One Menu
        --> 6 Generate All Menus for One Application
             7 Manage Libraries and Applications
             8 Previous Menu

                    Make your choice: 6

     NOTE: The meaning of function keys can change when running SQL*Forms
    ───
    v ^ Sun Feb 03 10:33:52 1991   BGM OSC DBG   Replace   DMU (DMUMNU)

Fig. 9.   Method of granting Work_Class assignment to new users of the SQL*Menu Application
PMAG used by the Global Paleomagnetic Database. (a) SQL*Menu Main Menu. Select Option 1.
(b) Menu Information Maintenance. Select Option 2. (c) Work_Class/User Information Form. Enter
new user under Work_Class 5 as described in the text. (d) Menu Information Maintenance. Select
Option 6 to update PMAG Menu library.

Work Class/User Information Form

Enter the Application_name PMAG and press [Enter]. The remainder of the Form showing Work_Class 5 will be displayed and you will note that the user WORKSHOP is already assigned to this Work_Class.

Press [Enter] until the cursor gets to the row below WORKSHOP and enter 5 for Work_Class then the userID under User_Name and N,Y,Y under the columns DBG,OSC,BGM just as for WORKSHOP.

Now press [F10] and this will commit this user to the Menu Application PMAG.

Press [Esc] or [Shift][F10] and you will return to the Menu Information Maintenance as shown in Figure 9(d).

Menu Information Maintenance

Select Option 6 – Generate All Menus for One Application
        You will be prompted for the application name, press [Enter] and a new Menu Library PMAGMNU.DMM will be created in the work directory C:\PALEOMAG. The new user will be permanently installed.

You will receive a number of messages that various tables are being updated and finally a message "press any key to return to menu" will be displayed.

You may now exit from SQL*Menu either with [Esc], or [Shift][F10], or by selecting Previous Menu until you return to the Application Menu. Then select Exit.

*Dropping Users from SQL*Menu*

To drop a user from an application or from SQL*Menu itself move to the C:\ORACLE5\DMU directory. Invoke SQL*Plus and run DELUSR as follows:

        C:\ORACLE5\DMU >SQLPLUS SYSTEM/Password
        SQL> START DELUSR

You will be prompted to enter the name of the application and the name of the user. The user's Work_Class Assignment will be removed from the specified application. Remember to EXIT from SQL*Plus.

*Reinstalling the SQL*Menu Application PMAG*

The Menu Application Tables are automatically imported into ORACLE when you import the GPMDB onto your system. If, for some reason, these tables get deleted it is still possible to reinstall the Application Menu PMAG using the special command file PMAGINS.SQL provided with the database and located with the customised files in directory C:\PALEOMAG. Before using PMAG-INS.SQL the Menu Application Tables must be reinstalled by ORACLE using the file MENUINS.BAT that resides in the C:\ORACLE5\BIN directory. The procedure is as follows and must be carried out in the C:ORACLE5\BIN directory:

> C:\ORACLE5\BIN >MENUINS SYSTEM password

When these tables are reinstalled, install the Application Menu PMAG in the Menu Tables as follows. Change to the C:\PALEOMAG directory log into SQL*Plus as SYSTEM/password and run PMAGINS by entering:

> C:\ORACLE5\BIN >CD \PALEOMAG
> C:\PALEOMAG >SQLPLUS SYSTEM/password
> SQL> START PMAGINS
> SQL> EXIT
> C:\PALEOMAG >

5.6. CHECKING DATABASE STORAGE

The database storage can be checked by entering the following command at the MS-DOS prompt:

> C:\PALEOMAG >CHKDBS SYSTEM systempassword

ORACLE displays
- the total size of the database files
- the amount of space available (i.e. not allocated to existing tables, views and indexes)

If the database users are creating large tables of their own eventually the database will give an "out of extents" error message when no more storage is available. You should regularly check the database storage and either EXPAND the database files (see Section III.5.1 – Creating a New Database File) or EXPORT tables that are no longer needed (see Section III.6.1 – The EXPORT Utility). Then DROP them from the database to release space (see Section V.6.2 – The DROP Table Command for a description of the DROP command).

5.7. EXPANDING THE DATABASE

The ORACLE database size is specified when it is initialised. This limits the amount of data that can be stored in the database. However if it outgrows the initial capacity it can be EXPANDed by adding a new database file. The simplest (but not the only) way to do this is to run the ORACLE utility EXPAND. This automatically creates a new database file (using the CCF utility) and adds the file to the SYSTEM partition (by issuing an ALTER PARTITION SYSTEM command). These two steps can be run separately but we have found that errors in issuing the commands are very easily made and correcting the havoc that results is time consuming. We recommend using the EXPAND command. The general syntax is:

C:\ORACLE5\DBS >EXPAND SYSTEM/password filename filesize

where

- filename is the name of a new file. The size of an existing file cannot be changed by the EXPAND command. This is the reason in Section III.5.1 (Creating a New Database File) that DBS2.ORA was deleted then re-created as a larger file.
- filesize is the size of the file in MS-DOS blocks. MS-DOS blocks are 512 bytes and ORACLE blocks are 1024 bytes. Therefore you must specify twice as many MS-DOS blocks as the number of ORACLE blocks required. For example to create a database file of 4 Megabytes enter:

EXPAND SYSTEM/password filename 8196

6. Backing Up and Restoring Your Database

6.1. THE EXPORT UTILITY

The EXPORT utillty provides a convenient way to backup your database, either as a whole, or any parts that you nominate. The data that have been EXPORTed can then be reinstalled into the ORACLE system using the IMPORT utility described in the next section.

Users (other than the DBA) can only EXPORT tables, views and other objects that they actually own (i.e. have created themselves). Only the DBA can EXPORT or IMPORT tables that belong to any user. Since the GPMDB has been provided as an EXPORT from the original creator called DATABASE only the DBA (SYSTEM) can IMPORT the GPMDB into your own system. The diskettes with the GPMDB thus become the ultimate backup for the entire GPMDB. This section will therefore concentrate on the requirements for individual users to backup their own data, or versions of the GPMDB, using EXPORT.

EXPORT writes data from the ORACLE database system to any specified target. This target may be a floppy diskette(s) or an MS-DOS file on your hard disk. There are three modes of EXPORT known as *Users Mode, Tables Mode* or *Full Database Mode*. Only the DBA can choose Full Database Mode. Table VI lists the items that can be EXPORTed for each of the three modes. Users will choose whether they wish to EXPORT in *Users Mode* or *Tables Mode*. Table VI shows that the *Users Mode* will EXPORT everything except views and synonyms, whereas the *Tables Mode* is rather more restrictive. In general one would use the *Tables Mode* just to backup a particular table of data and the *Users Mode* for a general backup of everything you own.

When users make changes to their tables the way in which ORACLE stores the table can be affected. If, for example, a lot of new entries are made to the table or many entries are updated or deleted, the tables will be stored in an increasingly

TABLE VI
List of items that can be EXPORTed in the various modes

	EXPORT Mode		
Item	Tables	User	Full Database
Table definitions	Y	Y	Y
Table data	Y	Y	Y
Space definitions	Y	Y	Y
Clusters	N	Y	Y
Indexes	N	Y	Y
First-level grants	N	Y	Y
All grants	N	N	Y
Views	N	N	Y
Synonyms	N	N	Y
Objects owned by SYS	N	N	N

fragmented way. This may slow data retrieval from the database quite dramatically. Answering YES to "Compress extents" during the EXPORT procedure removes this fragmentation and compresses the data into one large extent. On IMPORT the data will then be reinstalled in the database in the most efficient way.

The EXPORT command (EXP) is best carried out in the C:\PALEOMAG directory as follows:

C:\PALEOMAG >EXP userID/password

You will now be asked a series of questions:

Enter array fetch buffer size(default is 4096)>   [Enter]

Export file: EXPDAT.DMP > Enter the file name otherwise the default is used. Remember to use A:filename if a floppy diskette is being used.

U(sers), T(ables): U> Choose U or T for the Mode   [Enter]

Export Grants (Y/N): N> Y   [Enter]

Export the rows (Y/N): Y> Y   [Enter]

Compress extents (Y/N): Y> Y   [Enter]

If you chose *Tables Mode* you will not be prompted as to whether you wish to EXPORT the grants. In *Users Mode* EXPORT will then give the message:

Exporting userID
• Exporting user userID

- Exporting table Tablename   NNN Rows exported
  . . . . etc.

and so on until all the users tables have been EXPORTed. In *Tables Mode* you
will be prompted for the names of the Tables you wish to EXPORT with:

Exporting Specified Tables.
Table Name: > Enter the name of any table you own.   [Enter]

- Exporting table Tablename   NNN Rows exported
  . . . . . . etc.

   You will continue to be asked for new Tablenames. When completed press
[Enter] at the prompt for Table Name.
   Full details and some examples of EXPORT in various modes are given in the
'ORACLE Utilities User's Guide'.

6.2. THE IMPORT UTILITY

We have already described the procedure for IMPORTing the GPMDB from the
floppy diskettes containing the database. Table V of Section III.4.2 (Installing the
Global Paleomagnetic Database) gives an example of the full IMPORT procedure
with the prompts and messages that are received. Note that if you are reimporting
a table that already exists in the database you must first DROP the table from
the database. Otherwise the existing table will have all the rows duplicated and
will double the space requirement for the table in the database.
   The IMPORT (IMP) command is best carried out in your work directory
C:\PALEOMAG. You may restrict the items that are IMPORTed by answering
No to the message "Import of entire import file requested" as follows:

             C:\PALEOMAG >IMP userID/password

The following questions will be asked as in Table V.

Import file: EXPDAT.DMP > Enter the filename, remember to put A:filename
                         if using a floppy diskette.

Enter insert buffer size (default is 10240, min is 4096)>   [Enter]

List contents of import file only (Y/N): N > N   [Enter]

Ignore create errors due to object existence (Y/N): Y > Y: [Enter]

Import grants (Y/N): Y > Y   [Enter]

Import the rows (Y/N): Y > Y   [Enter]

Import of entire import file requested (Y/N): Y > N   [Enter]

Username: userID

At this stage you will be prompted for the names of the tables you wish to IMPORT.

Enter list of table names. Null list means all tables for user

Enter table name or . if done: Tablename1   [Enter]

Enter table name or . if done: Tablename2   [Enter]

Enter table name or . if done: .   [Enter]

. Importing user userID
. . Importing table "Tablename1"   NNN Rows imported
. . Importing table "Tablename2"   NNN Rows imported
. . . . .etc.

Examples of IMPORT sessions are provided in more detail in the 'ORACLE Utilities User's Guide'.

7. Factors Affecting ORACLE Performance

7.1. PC SYSTEM

The type of system on which you have installed the ORACLE RDBMS will have a marked effect on how quickly operations are carried out. If you have a 6 Megahertz 80286 system (as we do) performance will be considerably more pedestrian than if your system is say a 20 Megahertz 80486. However, there are ways to ensure that your system performance is not unnecessarily downgraded.

7.2. FRAGMENTATION

There are two possible sources of fragmentation. One is in the MS-DOS management system and the other is within ORACLE itself.

MS-DOS

During the preparation of your system for the installation of the GPMDB you were advised to pack (defragment) the hard disk. As part of normal hard disk usage, files and directories are created and deleted, with the result that disk space will become fragmented over time. ORACLE RDBMS database and BI files require a large amount of disk space, currently about 8 Megabytes for the database files and 4 Megabytes for the BI file. When the database and BI files are created, ORACLE will select the largest blocks of contiguous files space available on the disk. If all the space available is contiguous, ORACLE can assign space with greatest efficiency.

ORACLE

When tables are created ORACLE assigns space in the database files according to a SPACE definition. We have accepted the default SPACE definition that assigns an initial extent of 5 ORACLE blocks (each of 1024 bytes) to the table. When this is filled further extents in increments of 25 blocks are assigned. Twenty percent of each block is reserved for storing changes made to the data in the table. When the GPMDB is EXPORTed the tables are compressed into a single extent, as described in the previous section (Section III.6), to maximise the efficiency of table storage when the GPMDB is IMPORTed on to your system. If users create their own tables from the GPMDB tables, and then delete, change, or insert data into these tables the changes may require more space than the 20% reserved in each of the original blocks of storage. To provide this extra space, additional blocks will be chained to the original blocks. The larger the number of changes made to the data in the tables the greater the number of blocks that will need to be chained. Data in the tables will become non-sequential and storage within the ORACLE database tables will become fragmented. This greatly slows the rate at which data can be retrieved from the database.

To improve ORACLE performance carry out the following steps:

1. EXPORT your tables from the GPMDB, and compress extents to remove the chained blocks.
2. DROP your tables from the GPMDB – this can only be done by the owner of the tables.
3. IMPORT your compressed tables.

Refer to Section III.6 for details of the EXPORT and IMPORT procedures.

7.3. Protected versus real mode ORACLE performance

The 80286, 80386 and 80486 processor-based systems usually require more time to execute programs in P mode. To allow applications to run in P mode, ORACLE SQLPME must switch between the MS-DOS native real mode and protected mode. Due to performance limitations associated with the 80286 system, ORACLE tools, installed on this type of machine in P mode, will run significantly slower than the same tool installed in R mode. For 80386 and 80486 systems, performance reduction in P mode is not as significant as for 80286 systems.

7.4. ORACLE performance tuning

ORACLE INIT Parameters

The INIT.ORA file contains system parameters that are set each time the ORACLE RDBMS is started with the IOR command. The INIT.ORA file contains the System Global Area (SGA), the shared storage area in main memory that is the centre of data activity while ORACLE is running. Some SGA parameters can be adjusted to improve ORACLE performance. However, fine tuning of the SGA parameters will yield only a few percent improvement in ORACLE performance (about 5%). Tuning SQL code can improve performance by 2–300%. It is therefore worthwhile to spend the time to gain an understanding of indexes and writing SQL code to optimise the use of available indexes.

ORACLE Indexes

Indexes are small tables that ORACLE can use to locate data in the database without using a full table scan.

Indexes can:

- speed retrieval of data from the database especially from tables with a large number of rows (non-unique indexes).
- ensure each entry in a column or combination of columns of a table is unique and hence can uniquely identify rows in the table (unique indexes).

Unique and non-unique indexes may be defined on either a single column of a table or jointly on several columns of a table (concatenated index). A concatenated non-unique index should be created when several columns of a tables are almost always selected together in the same order. A concatenated unique index, defined on the entries of several columns of a table jointly uniquely identifies each row in the table.

Indexes may be defined as compressed or non-compressed. Compressed indexes are forward and rear compressed and take less database file storage space. Non-compressed indexes take up more space but may allow faster data retrieval than compressed indexes as values can be directly looked up in the index.

Once an index has been created, ORACLE automatically maintains it. When changes are made to the columns in a table the index is updated to incorporate the changes. The ORACLE optimiser determines when it should be used. The optimiser decides which way the data should be accessed and what indexes should be used to satisfy the SQL query.

An existing index will be used in a SQL query only if:

- the query contains a WHERE clause
- the indexed column is one used to include or exclude rows returned by a query
- the indexed columns are not modified by ORACLE functions or arithmetic operators

It is tempting to create indexes on all the columns of each table in the database, but indexes require database file storage space. Also, if the ORACLE optimiser has the choice of too many indexes query performance may be degraded. It is advisable to limit indexes to those that give worthwhile improvements in query performance. In general, the most appropriate columns of a table to index are those that are used frequently in data retrieval, especially primary keys and other

TABLE VII

List of indexes created for the Global Paleomagnetic Database. (The creator of both the tables and indexes is DATABASE)

TABLE NAME	COLUMNS INDEXED	INDEX NAME	INDEX TYPE	COMPRESS? (Y/N)
ALTRESULT	RESULTNO	CARRSLTI	UNIQUE	Y
AUTHORS	AUTHORS	NAUTHI	NON UNIQUE	N
	REFNO	CAREFI	UNIQUE	Y
CROSSREF	CATNO	CRRSLTCATI	UNIQUE	Y
	RESULTNO	CRRSLTCATI	UNIQUE	Y
FIELDTESTS	RESULTNO	CFTRSLTSTI	UNIQUE	Y
	TESTTYPE	CFTRSLTSTI	UNIQUE	Y
PMAGRESULT	HIMAGAGE	CPMRHIMI	NON UNIQUE	Y
	LOMAGAGE	CPMRLOMI	NON UNIQUE	Y
	RESULTNO	CPMRRSLTI	UNIQUE	Y
	ROCKUNITNO	CPMRROCKI	NON UNIQUE	Y
REFERENCE	REFNO	CREFREFI	UNIQUE	Y
REMARKS	REFNO	CREMREFI	UNIQUE	Y
ROCKUNIT	HIGHAGE	CRUHIAI	NON UNIQUE	Y
	LOWAGE	CRULOAI	NON UNIQUE	Y
	REFNO	CRUREFI	NON UNIQUE	Y
	ROCKUNITNO	CRUROCKI	UNIQUE	Y

columns used in joins, and those used in data sorts. A query that joins tables with indexed primary keys operates many times faster than the same query without indexes. The difference between the times depends on the complexity of the query and the number of tables joined. Usually the more tables to be joined the greater the improvement in performance. Unique indexes are more efficient than non-unique indexes as there are no duplicate entries. The ORACLE optimiser uses unique indexes in preference to non-unique indexes. Where multiple tables are joined, the unique indexes on the primary keys are used and the non-unique indexes on data selection conditions are not.

We have created unique indexes on the primary keys of each table in the GPMDB. After testing queries that might be issued to a single table in the GPMDB with indexes on various columns, we found few indexes gave significant query performance gains. This is because columns such as ROCKNAME, PLACE AND ROCKTYPE in the ROCKUNIT table, that may used in WHERE clauses or in data sorts, must be prefixed by '%' in the query. This automatically suppresses the use of an index. Indexes are not useful where a column contains only a few different values. The CONTINENT column of the ROCKUNIT table has only 12 distinct entries corresponding to the continents and oceans of the world, so an index does not improve performance. Table VII lists the indexes supplied with the GPMDB. For optimal GPMDB performance users should ensure their SQL commands are written to take advantage of these indexes. Section V.5.8 provides some guidelines for optimising SQL code.

IV. DATABASE QUERIES WITH SQL*MENU AND SQL*FORMS

1. SQL*Menu with the Global Paleomagnetic Database

1.1. STARTING SQL*MENU

A special Application Menu called PMAG has been developed for use with the GPMDB. This Menu was automatically imported into the ORACLE system when the GPMDB was first loaded.

SQL*Menu is slightly different from SQL*Plus and SQL*Forms in that all users must independently be granted access to SQL*Menu. Your DBA should arrange this access and must also grant you a 'Work-Class assignment' to the Application Menu PMAG. Details of this procedure are given in Section III. The GPMDB when imported into ORACLE has a general user WORKSHOP with Password PMAG automatically created. WORKSHOP also has automatic access to the Application Menu PMAG. So you may in the first instance make use of this access using WORKSHOP/PMAG.

The simplest way to make use of the GPMDB Menu is always to start ORACLE at every session by using the command:

C:\>LOADORA

This command may be given from any directory and will automatically load ORACLE into extended memory, place the user in the C:\PALEOMAG work directory and then load SQL*Menu. You will then be prompted for your UserID and Password as shown in Figure 10(a) and the Application Menu shown in Figure 10(b) will then be displayed. Select Option 1 and the GPMDB Main Menu will be displayed as in Figure 11(a).

If at any time during a session you exit from SQL*Menu, then ORACLE will still be loaded. To return to SQL*Menu you should not use the command LOADORA but instead in the C:\PALEOMAG directory enter the command:

C:\PALEOMAG > SQLMENU

You will again be prompted for your UserID and Password as before. After you have finished your session, exit from SQL*Menu and enter the command:

C:\PALEOMAG > SHUTORA

This will shut down ORACLE and remove it from RAM. Never try to use the command SHUTORA from within SQL*Menu!

1.2. THE MAIN MENU FOR THE GPMDB

*Option 1. SQL*Forms Options*

Selecting Option 1 from the Main Menu will display the SQL*Forms Option Menu

(a)

A U T H O R I Z A T I O N   F O R M

Enter username: WORKSHOP

Enter password:

Sun Feb 03 10:11:31 1991   Replace   page 1 : 1

(b)

A P P L I C A T I O N   M E N U

SELECT AN APPLICATION FROM THE LIST BELOW

--> 1   GLOBAL PALEOMAGNETIC DATABASE
     2   Exit

Make your choice: 1

---

v   Sun Feb 03 10:12:06 1991   Replace   (Sapl$)

Fig. 10.   Logging on to SQL*Menu application PMAG. (a) Authorization form to enter userID and
password. (b) The main application menu listing the GPMDB.

shown in Figure 11(b). Operating the Form PALEOMAG is described in detail
in the next section that deals with SQL*Forms. Options 2 to 5 in the SQL*Forms
Menu automatically call up the four special look-up tables. However these forms
can also be displayed directly from within the PALEOMAG Form.

(a)

```
G L O B A L    P A L E O M A G N E T I C    D A T A B A S E

                    MAIN MENU


        -->  1  SQL*Forms Options
             2  SQL*Plus Screen Options
             3  SQL*Plus Printer Options
             4  Oracle Tools and Utilities
             5  Previous Menu
             6  Exit




                Make your choice:  1

                 Gondwana Consultants
```

v	Sun Feb 03 10:40:06 1991	BGM OSC	Replace	PMAG (PMAG)

(b)

```
          S Q L * F o r m s    O p t i o n s

                -------------------


    -->  1  Use the Form PALEOMAG to Query the Whole Database
         2  Use the Form KEYS to check Function Keys for PALEOMAG Form
         3  Use the Form TIME to view the 1989 Geologic Time Scale
         4  Use the Form JOURNAL to see standard formats
         5  Use the Form INFO to check symbols
         6  Previous Menu
         7  Exit




                Make your choice:  1

          Press HELP (F2) for further information
```

v	Sun Feb 03 10:41:41 1991	BGM OSC	Replace	PMAG (FORMS)

Fig. 11(a, b).

Option 2. SQL*Plus Screen Options

Selecting Option 2 from the Main Menu will display the SQL*Plus Screen Options Menu shown in Figure 11(c). The Options listed run the various command files that have been prepared for use with the GPMDB. Details of these files are given in Section V.4.3 and are listed in Table XIV.

(c)
```
        S Q L * P l u s    S c r e e n    O p t i o n s

              -------------------------


-->   1   Basic Pole Data by Continent and Age - All magnetizations
      2   Basic Pole Data by Continent and Age - Primary magnetizations
      3   Extended Pole Data by Continent and Age - All magnetizations
      4   Extended Pole Data by Continent and Age - Primary magnetizations
      5   Reference List for a given Author
      6   Search for Reference - Author spelling and Year uncertain
      7   Listing of 1989 Geologic Time Scale
      8   Listing of Function Keys used in the PALEOMAG Form
      9   Listing of Explanation of Symbols used in the Global Database
     10   Previous Menu
     11   Exit

                    Make your choice:  1

             Press HELP (F2) for further information
```
```
v    Sun Feb 03 10:13:16 1991      BGM OSC         Replace      PMAG (SPLUS )
```

(d)
```
        S Q L * P l u s    P r i n t e r    O p t i o n s

              -----------------


-->   1   Basic Pole Data by Continent and Age - All magnetizations
      2   Basic Pole Data by Continent and Age - Primary magnetizations
      3   Extended Pole Data by Continent and Age - All magnetizations
      4   Extended Pole Data by Continent and Age - Primary magnetizations
      5   Reference List for a given Author
      6   Listing of the 1989 Geologic Time Scale
      7   Listing of Function Keys for use with PALEOMAG Form
      8   Listing of Explanation of Symbols used in the Global Database
      9   Listing of Journal styles used in the Global Database
     10   Previous Menu
     11   Exit

                    Make your choice:  1

             Press HELP (F2) for further information
```
```
v    Sun Feb 03 10:13:47 1991      BGM OSC         Replace      PMAG (PPLUS )
```
Fig. 11(c, d).

1. Basic Pole Data by Continent and Age – All Magnetizations

This option runs the command file SALLCAGE.SQL that is described in Table IX of Section V.4.3. You will be prompted for the name of a continent and an age range for which you want data using the prompts for 'loage' and 'hiage'. The output gives ROCKNAME, PLACE, TESTS, DEMAGCODE, PLAT, PLONG and DP, DM. The selection of data depends on the magnetic age of the magnetiz-

(e)

```
Oracle    Tools    and    Utilities
                  --------------------------

        -->  1  SQL*Plus
             2  Run  a  DOS  Command
             3  Run  a  FORM
             4  SQL*Forms  (Designer  only)
             5  Export
             6  Import
             7  Previous  Menu
             8  Exit



             Make  your  choice:  1

        Press  HELP  (F2)  for  further  information
```

| v | Sun Feb 03 10:38:14 1991 | BGH OSC | Replace | PMAG (ORATU) |

Fig. 11.   Global Paleomagnetic Database menus. (a) The main menu. (b) SQL*Forms options menu. (c) SQL*Plus screen options menu. (d) SQL*Plus printer options menu. (e) Oracle tool and utilities menu.

ation component. This means that all magnetizations whether primary, secondary etc. within the specified age range will be selected.

2. Basic Pole Data by Continent and Age – Primary Magnetizations

This option runs the command file SPRICAGE.SQL and gives the same output as Option 1 when the rock age is the same as the magnetization age. This does not guarantee that the magnetization components selected are primary, but at least all the known secondary magnetizations are omitted.

3. Extended Pole Data by Continent and Age – All magnetizations

This option runs the command file SRSLTALL.SQL and selects the same data as in Option 1 but in addition provides AUTHORS, YEAR, B (number of sites) and N (number of samples).

4. Extended Pole Data by Continent and Age – Primary Magnetizations

This option runs the command file SRSLTPRI.SQL and selects the same data as in Option 3 when the rock age and the magnetization age are identical. As with Option 2 this does not guarantee that only primary magnetizations are selected, but eliminates the known secondary magnetizations.

5. Reference List for a given Author

This option runs the command file SREFLIST.SQL and output is formatted like a standard reference list. You are prompted for the Authorname and need only provide a surname with leading capital. No initials are required.

6. Search for Reference – Author spelling and Year uncertain

This option runs the command file SAUTHQRY.SQL and provides the same reference style output as Option 5. This query is designed for the situation where you are uncertain of the correct spelling of the Authorname and uncertain of the exact year of a reference. When prompted for the authorname enter as much of the name as you can. It does not matter if the beginning, middle or end of the name is used. For example one could enter hinny if unsure how McElhinny was spelt. You are then prompted for the year. If uncertain, enter the decade by omitting the last digit of the year. For example for the decade of the 1980s enter 198 only.

The next three options list useful information from look-up tables. They can be printed directly using the Printer Options in the next Menu.

7. Listing of the 1989 Geologic Time Scale

8. Listing of Function Keys used in the PALEOMAG Form

9. Listing of Explanation of Symbols used in the Global Database

Option 3. SQL*Plus Printer Options

Selecting Option 3 from the Main Menu will give the SQL*Plus Printer Options Menu shown in Figure 11(d), which is virtually the same as that for the Screen Options. Output is automatically sent to the printer. Make sure that your printer is switched on before using any of the printer options.

Option 4. Oracle Tools and Utilities

Selecting Option 4 from the the Main Menu will give the Tools and Utilities Options Menu shown in Figure 11(e). This Menu is similar to that provided with the utility Client Manager, which is why we suggested it was unnecessary to load that utility.

Option 1 will put you directly into SQL*Plus using the UserID and Password from SQL*Menu. To return to the Menu use the command EXIT at the SQL prompt.

Option 2 will place you in MS-DOS with the symbol OS (operating system) before the prompt. You can issue any MS-DOS commands as though you were in the PALEOMAG work directory. To return to the Menu press [Enter] at the prompt.

Option 3 enables you to run a Form independently of the SQL*Forms Menu described above. You will be prompted for the name of the Form when this Option is selected. To return to the Menu from any Form press [Esc] or [Shift][F10].

Option 4 can only be used if the SQL*Forms designer building tables have been installed. See details under SQL*Forms later in this section.

Options 5 and 6 allow you to run the Export and Import utilities directly from SQL*Menu. Their operation is described in Section III.6 (Backing Up and Restoring Your Database).

2. Running SQL*Forms (RUNFORM)

2.1. ELEMENTS OF A FORM

A Form is composed of one or more PAGES. Each page contains one or more BLOCKS, and is equivalent to one screen display. Navigation around the Form is by BLOCKS, RECORDS and FIELDS, not by pages.

A PAGE is a portion of a Form that is displayed on one screen.
A BLOCK is based on one of the TABLES in the database.
A RECORD is one ROW in a table in the database.
A FIELD is one entry in a COLUMN in a table in the database.

A typical SQL*Forms screen (see example for the customised Form PALEO-MAG in Figure 12) displays a block title at the top of the screen. Below the title are fields that can display data values. At the bottom of the screen are the message and status lines. The message line displays SQL*Forms messages. Prompts are issued on the status line. The status line may contain:

Char Mode:	Either Replace or Insert will be displayed.
Page N	The page number (N) currently displayed.
ENTER QUERY	You have pressed [Enter Query] key and have not yet pressed [Execute Query] or [Exit/Cancel] keys.
Count: N	The number of the record (N) retrieved by a query. Each time you display a record fetched by a query, the count is increased. When the last record has been fetched an asterisk * is displayed before the count.

It is highly recommended that you work through the step-by-step tutorial in the 'SQL*Forms Operator's Guide', before attempting to run the custom designed Form supplied with the database.

2.2. CUSTOMBUILT FORM – PALEOMAG

The customised Form supplied with the database is called PALEOMAG and the code (PALEOMAG.FRM) is stored in the C:\PALEOMAG directory. Before running a Form make sure ORACLE is loaded. The PALEOMAG Form enables

```
(a)
+-------------------------------========--AUTHORS--=========-----------------------+
|                                                                                  |
|   REFNO   _____                                                                 |
|                                                                                  |
|   AUTHORS  _____ |
|                                                                                  |
|                            REFERENCED BY OR                                      |
|                            NO ROCKS RESULTS                                       |
|                            _____                                       |
|                                                                                  |
+-----------------------========--REFERENCE--========------------------------------+
|                                                                                  |
|   REFNO  _____    YEAR  _____                                                |
|                                                                                  |
|   JOURNAL  _____|
|                                                                                  |
|   VOLUME  _____    PAGES  _____                                          |
|                                                                                  |
|   TITLE  _____|
|                                                                                  |
+----------------------------------------------------------------------------------+
Press [F2] for list of Function Keys for this Form. Press [Esc] to Exit.
        Char Mode: Replace   Page 1                    Count: *0


+------------------------=========--ROCKUNIT--=========----------------------------+
|                                                                                  |
|      REFNO  _____    ROCKUNITNO  _____     RLAT  _____   RLONG  _____       |
|                                                                                  |
|   ROCKNAME  _____|
|                                                                                  |
|      PLACE  _____|
|                                                                                  |
|  CONTINENT  _____        TERRANE  _____|
|                                                                                  |
|   ROCKTYPE  _____                  |
|                                                                                  |
|     STRATA  _____|
|                                                                                  |
|   STRATAGE  _____        LATSPREAD  _____                          |
|                                                                                  |
|     LOWAGE  _____  HIGHAGE  _____   METHOD  _____|
|                                                                                  |
| ISOTOPEDATA  _____|
|                                                                                  |
|  STRUCTURE  _____|
+----------------------------------------------------------------------------------+
Press [F2]-Function Keys, [F4]-Time Scale, [F5]-Symbols. Exit with [Esc].
        Char Mode: Replace   Page 2                    Count: *0
```

Fig. 12a.

you to query the complete database, that is every table and column in the database. The Form consists of blocks with the same name as the table on which the block is based. Each field in the block corresponds to a column in the table. The PALEOMAG Form comprises 4 pages:

Page 1   AUTHORS block, REFERENCE block
Page 2   ROCKUNIT block
Page 3   PMAGRESULT block
Page 4   ALTRESULT block, FIELDTESTS (multi-record) block, CROSSREF (multi-record) block

Figure 12 shows the details of each page of the Form. A block diagram of the

(b)

```
+------------------------------========--PMAGRESULT--=========---------------------+
|   ROCK    RESULT                      LO       HI                                 |
|   UNITNO    NO       COMPONENT      MAGAGE   MAGAGE        TESTS          TILT     |
|                                                                                   |
|   ____    ____       ____         ____     ____         ____          ____        |
|                                                                                   |
|   SLAT    SLONG        B            N         DEC      INC           KD      ED95  |
|                                                                                   |
|   ____    ____   P   ____         ____    NO    ANTI-  ____          ____    ____  |
|   PLAT    PLONG TYPE  DP           DM   REVERSED  PODAL    N_NORM  D_NORM  I_NORM  |
|                                                                                   |
|   ____    ____  __   ____         ____    ____   ____     ____    DEMAG   TREAT    |
|   K_NORM  ED_NORM     N_REV       D_REV   I_REV   K_REV   ED_REV   CODE   -MENT    |
|                                                                                   |
|   ____    ____        ____        ____    ____   ____     ____    ____    ____     |
|   LABDETAILS    _____ |
|   ROCKMAG       _____ |
|                                                                                   |
|   N_TILT   D_UNCOR I_UNCOR   K1     ED1       D_COR   I_COR      K2      ED2       |
|                                                                                   |
|   ____    ____    ____     ____    ____      ____    ____      ____    ____        |
|   COMMENTS      _____ |
+-----------------------------------------------------------------------------------+
Press [F2]-Function Keys, [F4]-Time Scale, [F5]-Symbols.   Exit with [Esc].
         Char Mode: Replace   Page 3                       Count: *0


+------------------------------=========--ALTRESULT--========---------------------+
|   RESULT                                                                         |
|    NO                   APLAT     APLONG                   KP          EP95      |
|                                                                                 |
|   ____                  ____      ____                    ____         ____      |
+---------------------------------========--FIELDTESTS--========-------------------+
|   RESULT TEST                                                          SIGNIF    |
|    NO    TYPE PARAMETERS                                               ICANCE    |
|                                                                                 |
|   ____   __   _____   ____      |
|   ____   __   _____   ____      |
|   ____   __   _____   ____      |
+---------------------------========--CROSSREF--========--------------------------+
|   RESULT                                                                        |
|    NO      CATNO                                                                |
|                                                                                 |
|   ____    _____                                                          |
|   ____    _____                                                          |
|   ____    _____                                                          |
+---------------------------------------------------------------------------------+
Press [F2]-Function Keys, [F4]-Time Scale, [F5]-Symbols.   Exit with [Esc].
         Char Mode: Replace   Page 4                       Count: *0
```

Fig. 12. The 4 pages of the custombuilt form PALEOMAG.

Form layout with links to the various tables is shown in Figure 13. In addition to the tables in the GPMDB, two additional tables, named REMARKS and KEYS, are specifically used to assist with operation of the PALEOMAG Form. KEYS is a table that specifies the purpose of each of the Function Keys used in the PALEOMAG Form. REMARKS provides additional information regarding the number of records to be found on subsequent pages of the Form.

Running the Form

To run the PALEOMAG Form, ORACLE must first be loaded. Users of SQL* Menu can simply run the Form by selecting Option 1 from the Main Menu and then Option 1 from the SQL*Forms Options as described in Section 1.2 above.

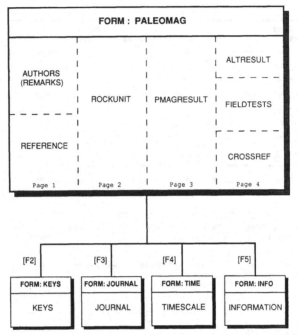

Fig. 13.   Block diagram of the form PALEOMAG showing linkages to accessory forms derived from various subsidiary tables.

Otherwise, after loading ORACLE enter the RUNFORM command in the C:\PA-LEOMAG work directory:

C:\PALEOMAG > RUNFORM PALEOMAG userID/password

The first page of the PALEOMAG Form (Figure 12) has a field labelled REFER-ENCED BY OR NO ROCKS, RESULTS. This is the REMARKS field in the AUTHORS block. In the figure the lines after, or under, the column names represent high-lighted fields where data are displayed, or queries are entered. On the screen, the Form will be in colour provided of course you have a colour monitor.

Note that the message at the bottom of the screen indicates that the function key [F2] will retrieve the KEYS table which will be displayed on another Form. All the four tables KEYS, JOURNAL, TIMESCALE and INFORMATION can be retrieved very simply directly from the PALEOMAG Form using the function keys [F2] through [F5] as listed in Table VIII. This can be done from any of the blocks of the Form.

Navigating Around the Form

You can navigate about the Form and query the database from the Form via pre-

TABLE VIII
Main function keys used in the PALEOMAG form

Operators Function	PC Key
Count query hits	[Shift][F2]
Enter query	[F7]
Execute query	[F8]
Exit/cancel	[Shift][F10] or [Esc]
Function keys (PALEOMAG)	[F2]
Journal styles	[F3]
Next block	[Ctrl][Page Down]
Next field	[Enter] or [Tab]
Next record	[Page Down]
Previous block	[Ctrl][Page Up]
Previous field	[Shift][Tab]
Previous record	[Page Up]
Retrieve matching records	[F6] (forward direction)
Retrieve matching records	[Shift][F6] (reverse direction)
Information about symbols	[F5]
Time scale	[F4]
Special retrieve key	[F9] (first REFERENCE from AUTHORS or FIELDTESTS from PMAGRESULT)

Other less often used keys

Clear block	[Shift][F5]
Clear field	[Ctrl][End]
Clear record	[Shift][F4]
Move down	[Down Arrow]
Move left	[Left Arrow]
Move right	[Right Arrow]
Move up	[Up Arrow]
Next set of records	[Ctrl][N]
Print	[Shift][F8]
Scroll left	[Ctrl][Left Arrow]
Scroll right	[Ctrl][Right Arrow]
Show function keys	[F1]

defined function keys. The definition of these varies with different terminal types and according to the .CRT definition chosen. The most common key definitions are listed in Table VIII and apply to the .CRT definitions BIOS and BIOSBLUE. The complete set of these key definitions is given in the KEYS table.

First try navigating backwards and forwards through successive blocks (and pages) of the Form using the keys

[Ctrl][Page Down] – moves the cursor to the first field of successive blocks in the forward direction

[Ctrl][Page Up] – moves the cursor to the first field of preceding blocks in the reverse direction

In the last block of the Form (CROSSREF) moving forward will take you to

```
(a)
+------------------------========--AUTHORS--=========----------------------+
|                                                                          |
|    REFNO _____                                                         |
|                                                                          |
|   AUTHORS %McElhinny%                                                     |
|                                                                          |
|                          REFERENCED BY OR                                |
|                          NO ROCKS,RESULTS                                |
|                          _____                                |
|                                                                          |
+-----------------------========--REFERENCE--=========---------------------+
|                                                                          |
|    REFNO _____     YEAR _____                                        |
|                                                                          |
|   JOURNAL _____|
|                                                                          |
|   VOLUME _____   PAGES _____                                     |
|                                                                          |
|    TITLE _____|
|                                                                          |
+--------------------------------------------------------------------------+
Enter a query; press F8 to execute, Shift-F10 to cancel.
         Char Mode: Replace  Page 1    ENTER QUERY     Count: *0
(b)
+------------------------========--AUTHORS--=========----------------------+
|                                                                          |
|    REFNO   1006                                                          |
|                                                                          |
|   AUTHORS McElhinny,M.W., Gough,D.I.                                     |
|                                                                          |
|                          REFERENCED BY OR                                |
|                          NO ROCKS,RESULTS                                |
|                          1,1 Results                                     |
|                                                                          |
+-----------------------========--REFERENCE--=========---------------------+
|                                                                          |
|    REFNO   1006     YEAR   1963                                          |
|                                                                          |
|   JOURNAL Geophys.J.Roy.Astron.Soc.                                      |
|                                                                          |
|   VOLUME    7   PAGES 287-303                                            |
|                                                                          |
|    TITLE The palaeomagnetism of the Great Dyke of Southern Rhodesia      |
|                                                                          |
+--------------------------------------------------------------------------+
Press [F2] for list of Function Keys for this Form. Press [Esc] to Exit.
         Char Mode: Replace  Page 1                    Count: *34
```

Fig. 14.   Example of a query using the name of an author. (a) Enter query [F7]. Place author name between %....% (b) The last record retrieved by the query in (a).

the first block again. In the first block (AUTHORS) moving backwards will take you to the last block.

Executing a Query

To execute a query in any block the cursor must first be positioned in a field within the block. For example, suppose you wish to find all the references for which McElhlnny was one of the authors. Enter and execute the query in the AUTHORS block as follows – see Figure 14(a):

Press [F7]   – the enter query key.

Press [Enter] – position the cursor in the AUTHORS fleld and type %McElhinny%. The % symbol stands for any string of characters, including none, so that McElhinny will be selected from any author list wherever the name occurs. Initials do not need to be known.

Press [F8]   – the execute query key.

This will retrieve all the references (list of authors only) in the database for which McElhinny was an author on a paper. There will be more than one record retrieved from the database as a result of this query. Underneath the list of authors is the REMARKS field providing additional information about each reference. The number of rockunits and results that are in subsequent blocks, are shown below NO ROCKS, RESULTS. If there are no results at all, then this reference has been referred to in another reference indicated by the given REFNO.

To view the first REFERENCE retrieved by the above query:

Press [F9] – the special retrieve key. This will automatically retrieve the matching details in the REFERENCE block.

If you wish to view the complete title of the reference:

Press [Ctrl][Page Down]   – to enter the REFERENCE block.
Press [Enter]   – to skip to the TITLE field.
Press [Ctrl][Right Arrow] – to scroll to the end of the title.

The title field is a maximum of 200 characters long. However a Form will permit a maximum display of 60 characters, so you must scroll the title to view all of it.

Press [Ctrl][Page Up] – to return to the AUTHORS block.

You may now page down the records that have been retrieved using the key:

[Page Down] – the next record key

Continue until all the records have been seen and Count: *N appears at the base of the Form. Note that all the matching reference details are automatically retrieved in REFERENCE.

You can reverse the process and all AUTHORS and REFERENCE details will be displayed in reverse order using the key:

[Page Up] – the previous record key

A message "At first record" is displayed when you have returned to the first record.

Figure 14(b) shows the last record retrieved by the above query.

Counting the Records to be Retrieved

It is sometimes convenient to determine in advance how many records will be retrieved by a query. To do this in the above case:

Press [F7]   – the enter query key. Move to the field you wish to query and
                    type in the query (e.g. type %McElhinny% in the
                    AUTHORS field of the AUTHORS block).
Press [Shift][F2] – the count query hits key.

This will give the number of records the query will return with the message "Query
wlll retrieve N records".

Now either execute the query with [F8] or decide that too many records will be
retrieved and cancel it with:

[Esc] or [Shift][F10] – the cancel query key.

Retrieving Matching Records from Succeeding Blocks

The PALEOMAG Form has been automated so that you can retrieve the matching
records progressively down the blocks of the Form. We saw above how the
matching REFERENCE was automatically retrieved for each AUTHORS record.
To demonstrate this we can use REFNO 1830 to illustrate a number of the features
of the database.

In the AUTHORS block:

Press [F7] – the enter query key. Type 1830 in the REFNO field.
Press [F8] – the execute query key.
Press [F9] – the special retrieve key, to obtain the matching REFERENCE as
             above (Figure 15(a), top).
Press [F6] – the retrieve matching records key.

All records in the ROCKUNIT block corresponding to this reference will be
retrieved (Figure 15(a), bottom).
    Count: *1 indicates that only one rock unit has been studied.

Press [F6] – the corresponding records will automatically be retrieved from the
             PMAGRESULT block (first one in Figure 15(b)).

Count: 1 indicates that there is more than one result. Note that in our example
this is a high temperature component, so we might also expect a low temperature
component.

Press [Page Down] – to view that record.

Count: *2 indicates this is the last record.

Press [Page Up] – to return to the first record.

The entries (C*+,F+) in the TESTS field indicate that there are some records in
the FIELDTESTS block.

Press [F9]   – the special retrieve key, to view these tests on the next page

(a)

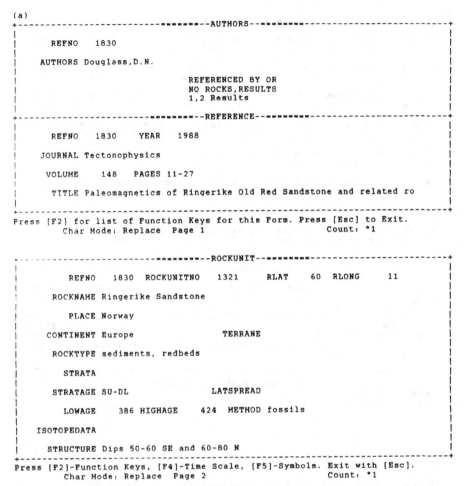

Fig. 15a.

of the Form in the FIELDTESTS block (Figure 15(b)). Two
TESTTYPEs (C*, F) are displayed for this result.

Press [Shift][F6] – the return key, to return to PMAGRESULT.

Press [F6]   – to query the ALTRESULT block for any matching record.

A message "Query caused no records to be retrieved. Re-enter" appears on the
message line at the bottom of the screen.

Press [F6]   – to query the FIELDTESTS block for any matching records.
                  The two TESTTYPESs shown in Figure 15(b) will again be
                  displayed.

Press [F6]   – to query the CROSSREF block for any matching records.

```
(b)
+----------------------------PMAGRESULT--======---------------------------+
|   ROCK    RESULT                     LO      HI                         |
|   UNITNO    NO      COMPONENT      MAGAGE  MAGAGE      TESTS        TILT |
|   1321     1556   High temp.        386     424    C*+,F+,Rc,M     100  |
|                                                                         |
|   SLAT    SLONG       B        N          DEC     INC       KD     ED95 |
|    60      11        19        67         26     -16       14.7    9.1  |
|                     P                     NO    ANTI-                    |
|   PLAT    PLONG  TYPE    DP     DM  REVERSED  PODAL   N_NORM  D_NORM I_NORM |
|    19      164    D    4.8    9.3  37        165.4     12       30   -20 |
|                                                            DEMAG   TREAT |
|   K_NORM  ED_NORM     N_REV   D_REV I_REV  K_REV  ED_REV    CODE   -MENT |
|   11.7     13.2         7      198    11   40.5    9.6       4       T   |
|                                                                         |
|   LABDETAILS   Component defined between 620-700C                       |
|   ROCKMAG      IRM(no sat.0.35T)                                        |
|                                                                         |
|   N_TILT   D_UNCOR I_UNCOR   K1    ED1     D_COR  I_COR    K2      ED2   |
|    19       32.9   -3.3     6.9   13.9      26    -16     14.7     9.1   |
|                                                                         |
|   COMMENTS                                                              |
+-------------------------------------------------------------------------+
Press [F2]-Function Keys, [F4]-Time Scale, [F5]-Symbols.  Exit with [Esc].
          Char Mode: Replace  Page 3                      Count: 1


+--------------------------=====----ALTRESULT--======---------------------+
|   RESULT                                                                |
|    NO                   APLAT     APLONG                KP       EP95    |
|   ____                  ____      ____                 ____      ____    |
+--------------------------=======--FIELDTESTS--========------------------+
|   RESULT TEST                                                  SIGNIF   |
|    NO   TYPE  PARAMETERS                                       ICANCE   |
|   1556   C*   Permo-Carboniferous dyke has overprinted rocks near margi  YES |
|   1556   F    k2/k1=2.15                                         95     |
|    ____                                                         ___     |
+-------------------------========-CROSSREF--========---------------------+
|   RESULT                                                                |
|    NO       CATNO                                                       |
|   ____      ____                                                        |
|   ____      ____                                                        |
|   ____      ____                                                        |
+-------------------------------------------------------------------------+
Press [F2]-Function Keys, [F4]-Time Scale, [F5]-Symbols.  Exit with [Esc].
          Char Mode: Replace  Page 4                      Count: *2
```

Fig. 15.   Query for Records from REFNO 1830 on the PALEOMAG Form as described in the text.
(a) Record for REFNO 1830 showing pages 1 and 2 of the Form. (b) Pages 3 and 4 of the Form
showing the High Temp. Component and corresponding Fieldtest records.

A message "Query caused no records to be retrieved. Re-enter" appears on the
message line at the bottom of the screen.

Press [Shift][F6] – the return key, and the cursor will return to the ROCKUN-
ITNO field of the PMAGRESULT block.

Retrieving Matching Records from Preceding Blocks

Use the navigation keys [Ctrl][Page Up] or [Ctrl][Page Down] to place the cursor
in the REFERENCE block on page 1 of the Form. Suppose you wish to obtain
all the references in the database for the year 1969 from the Journal of Geophysical

```
+----------------------------=======--AUTHORS--=========-----------------------+
|                                                                              |
|    REFNO _____                                                            |
|                                                                              |
|    AUTHORS _____    |
|                             REFERENCED BY OR                                 |
|                             NO ROCKS,RESULTS                                 |
|                             _____                                   |
|                                                                              |
+----------------------------=========--REFERENCE--=========-------------------+
|                                                                              |
|    REFNO _____      YEAR 1969                                             |
|                                                                              |
|    JOURNAL J.Geophys.Res.                                                    |
|                                                                              |
|    VOLUME _____      PAGES _____                                    |
|                                                                              |
|    TITLE _____       |
|                                                                              |
+------------------------------------------------------------------------------+
Enter a query;  press F8 to execute, Shift-F10 to cancel.
            Char Mode: Replace   Page 1      ENTER QUERY        Count: *0
```

```
+----------------------------=======--AUTHORS--=========-----------------------+
|                                                                              |
|    REFNO   479                                                               |
|                                                                              |
|    AUTHORS Faure,G., Chaudhuri,S., Fenton,M.D.                               |
|                             REFERENCED BY OR                                 |
|                             NO ROCKS,RESULTS                                 |
|                             REFNO 599,548                                    |
|                                                                              |
+----------------------------=========--REFERENCE--=========-------------------+
|                                                                              |
|    REFNO   479    YEAR   1969                                                |
|                                                                              |
|    JOURNAL J.Geophys.Res.                                                    |
|                                                                              |
|    VOLUME   74  PAGES 720-725                                                 |
|                                                                              |
|    TITLE Ages of the Duluth gabbro complex and of the Endion sill, Du        |
|                                                                              |
+------------------------------------------------------------------------------+
Press [F2] for Function Keys: Press F[3] for Journal Styles: Exit with [Esc].
            Char Mode: Replace   Page 1                      Count:  1
```

Fig. 16.   Query for records for the year 1969 from the *Journal of Geophysical Research*. The first record retrieved is displayed below.

Research. Note that the message line at the bottom of the screen reminds you that [F3] will provide a list of Journal styles used in the database.

Press [F3]   – call up the JOURNAL table and Form.

Under abbreviation 'JGR' the style is listed as 'J.Geophys.Res.'

Press [Esc]   – to return to the main Form and REFERENCE block again.
Press [F7]   – the enter query key.
Press [Enter] – to move to the YEAR field – type 1969.
Press [Enter] – to move to the JOURNAL field – type J.Geophys.Res.
Press [F8]   – the execute query key (Figure 16(a)).

The first record will be displayed, but we need the matching set of AUTHORS

```
+----------------------------========--ROCKUNIT--=========----------------------+
|                                                                                |
|       REFNO  _____   ROCKUNITNO _____      RLAT _____   RLONG _____     |
|                                                                                |
|     ROCKNAME  _____        |
|                                                                                |
|        PLACE  _____        |
|                                                                                |
|    CONTINENT  North America          TERRANE _____   |
|                                                                                |
|     ROCKTYPE  _____                    |
|                                                                                |
|       STRATA  _____        |
|                                                                                |
|     STRATAGE  _____              LATSPREAD _____                  |
|                                                                                |
|       LOWAGE  <146    HIGHAGE >65    METHOD _____    |
|                                                                                |
|   ISOTOPEDATA  _____         |
|                                                                                |
|     STRUCTURE  _____         |
+--------------------------------------------------------------------------------+
Query will retrieve 112 records.
        Char Mode: Replace  Page 2    ENTER QUERY       Count: *0
```

Fig. 17.   Query for all rockunits for North America whose ages overlap into the Cretaceous (between 65 and 146 Ma).

in the preceding block.

Press [Shift][F6] – the reverse retrieval key.

Now the corresponding AUTHORS record is displayed with the information REFNO 599,548 in the REMARKS field. This indicates that this reference is a cross-reference that is referred to in REFNO 599 and 548, and there are no rockunits or results associated with this reference (Figure 16(b)).

You can [Page Down] and [Page Up] to see all the records retrieved by the query and this will automatically retrieve all the matching AUTHORS in turn.

We can now try one or two more sophisticated queries in the ROCKUNIT block. Suppose you would like to obtain all the rock units from North America that have ages within the Cretaceous (65–146 Ma). Navigate to the ROCKUNIT block with [Ctrl][Page Up] or [Ctrl][Page Down] as necessary:

Press [F7]   – the enter query key.
Press [Enter]   – position the cursor in the CONTINENT field and type in North America.
Press [Enter]   – position the cursor in the LOWAGE field and type in <146.
Press [Enter]   – position the cursor in the HIGHAGE field and type in >65.
Press [Shift][F2] – to find out how many records will be retrieved (Figure 17).
Press [F8]   – to execute the query.

This query will produce a 'window' of all rock units in the age range 65–146 Ma. Do not enter LOWAGE >65 and HIGHAGE <146 as this excludes some of the data you are seeking. LOWAGE <146 and HIGHAGE >65 will retrieve results such as when LOWAGE is 84 and HIGHAGE is 212.

You may now retrieve the AUTHORS and REFERENCE for any rockunit displayed:

Press [Shift][F6] – the reverse retrieval key.

To return to the ROCKUNIT block:

Press [F6] – the forward retrieval key.

Making Queries Using Variables

More sophisticated queries can be made using placeholders by placing a variable in one or more fields and then completing a WHERE? clause at the bottom of the screen. Suppose you would like to find out which rock units have been studied within a nominated latitude and longitude grid, for example between 45.0 and 48.0 latitude and between −120.0 and −110.0 longitude.

In the ROCKUNIT block:

Press [F7]   – to enter the query.

Press [Enter] – position the cursor in the RLAT field and type in :R (to indicate the variable R is to be used here).

Press [Enter] – position the cursor in the RLONG field and type in :L (the variable L is to be used here).

:R and :L are *placeholders*, that is variables that have been used to define the fields RFLAT and RLONG respectively.

Press [F8]   – to execute the query. WHERE? will appear at the bottom of the screen with the cursor positioned next to it.

Type in the following:

WHERE? :R BETWEEN 45.0 AND 48.0 AND :L BETWEEN −120.0 AND −110.0

The screen is displayed in Figure 18.

Press [Enter] – to retrieve the records.

Note: BETWEEN must list the values from smallest to largest, hence −120.0 (smallest) must be listed before −110.0 (largest) If these are inverted, no records will be retrieved.

To view the other records retrieved by this query use [Page Down] and [Page Up] as usual.

Retrieving Entries Corresponding to Other Catalogue Numbers

The following query uses joins of all the tables in the database from the CROSSREF block. Suppose you want to find the records that correspond to a

```
+-------------------------========--ROCKUNIT--=========--------------------+
|                                                                          |
|     REFNO _____   ROCKUNITNO _____     RLAT :R    RLONG :L             |
|                                                                          |
|   ROCKNAME _____  |
|                                                                          |
|      PLACE _____  |
|                                                                          |
|  CONTINENT _____    TERRANE _____  |
|                                                                          |
|   ROCKTYPE _____              |
|                                                                          |
|     STRATA _____  |
|                                                                          |
|   STRATAGE _____        LATSPREAD _____                      |
|                                                                          |
|     LOWAGE _____ HIGHAGE _____  METHOD _____  |
|                                                                          |
| ISOTOPEDATA _____  |
|                                                                          |
|  STRUCTURE _____  |
+--------------------------------------------------------------------------+
```

Query Where... ? :R BETWEEN 45.0 AND 48.0 AND :L BETWEEN -120.0 AND -110.0

Fig. 18.   Using variables to query the PALEOMAG Form.

particular entry in one of the old paleomagnetic catalogues such as the Pole Lists of the Geophysical Journal. Navigate to the CROSSREF block using [Ctrl][Page Up] or [Ctrl][Page Down] as necessary:

Press [F7] – to enter a query. Move the cursor to the CATNO field and enter GJ 15.213 corresponding to entry number 213 in Pole List 15.
Press [F8] – to execute the query and retrieve the matching record that gives the RESULTNO in the database.
Press [F6] – the forward retrieval key to retrieve the corresponding AUTHORS and REFERENCE.

Now you are in a position to query down the successive blocks using the [F6] retrieve key as explained above.

NOTE: If at any time you receive the following message in the command line at the bottom of the screen,

"Do you wish to commit the changes you have made ?:Y"

you should respond N [Enter] as you are not permitted to make changes to the database. If you respond Y [Enter] you will receive an ORACLE error. You will not be able to exit from the Form until you respond N. This message occurs in some of the special situations following use of the [Shift][F6] key either in REFERENCES or PMAGRESULT or the [F6] key from CROSSREF.

Press [Esc] or [Shift][F10] – To exit from the Form and return to MS-DOS or SQL*Menu as the case may be.

Authors: **%McElhinny%** ....

Year: .... Journal: ....

Vol: .... Pages: ....

Title: ....

Rockunit: ....

Place: ....

Slat: .... Slon: .... L_age: .... H_age: ....

N: .... Dec: .... Inc: .... a95: ....

Plat: .... Plon: .... dp: .... dm: ....

Fig. 19a.

3. SQL*Forms Designer (SQLFORMS)

Any user may design and build a Form for his or her own purposes. However, this is not recommended until you become an experienced user of ORACLE. Refer to the appropriate ORACLE manuals for details.

To design and build Forms the SQL*Forms Extended Data Dictionary (a special set of tables) must be loaded into the ORACLE system. This may have been carried out on initial installation but the tables are dropped when the database was reinitialized before importing the GPMDB. Your DBA should arrange for these tables to be installed as described in Section II.2.3 (Using SQL*Forms under MS-DOS). You cannot design and build Forms with only 1 Mb of Extended memory over and above MS-DOS as there will be insufficient RAM for the purpose. You should have the full 2 Mb as we recommend.

Users should also arrange for their DBA to install the demonstration tables for the SQL*Forms Tutorial and the SQL*Forms Designer's Tutorial. It is recommended that you first work through the 'SQL*Forms Operator's Guide' supplied with ORACLE. This guide demonstrates the use of a sample Form (called SAMPLE) that is designed for use with a sample database that can be set up by executing the DEMOLBLD command file. As a second step you should then work through the 'SQL*Forms Designer's Tutorial'. The 'SQL*Forms Designer's Reference' contains all the information regarding the complexities available for Form design.

It is worth pointing out some of the pitfalls that can occur when naming Forms. Since SQL*Forms stores Forms in MS-DOS files, Form names must consist only of characters that are valid in MS- DOS file names. Note the following rules:

Authors: <u>McFadden,P.L., Ma,X.H., McElhinny,M.W., Zhang,Z.K.</u>

Year: <u>1988</u>   Journal: <u>Earth Planet.Sci.Letters</u>

Vol: <u>87</u>   Pages: <u>152-160</u>

Title: <u>Permo-Triassic magnetostratigraphy in China: northern Tarim</u>

Rockunit: <u>Redbeds, Tarim Basin</u>

Place: <u>China</u>

Slat: <u>42.1</u>	Slon: <u>83.3</u>	L_age: <u>245</u>	H_age: <u>255</u>
N: <u>103</u>	Dec: <u>201.8</u>	Inc: <u>-54.9</u>	a95: <u>5.5</u>
Plat: <u>71.8</u>	Plon: <u>187.6</u>	dp: <u>5.5</u>	dm: <u>7.8</u>

Fig. 19. Example of the GPMDB using the Macintosh HyperCard. (a) Query using the name of an author. (b) The first record retrieved of a total of 40.

Authors:

Year: <u>>=1980</u>   Journal:

Vol: <u>*</u>   Pages:

Title:

Rockunit:

Place: <u>%China%</u>

Slat:	Slon:	L_age:	H_age:
N:	Dec:	Inc:	a95:
Plat:	Plon:	dp:	dm:

Fig. 20a.

- A name should not consist of more than 8 characters or contain embedded periods
- If a name is longer than 8 characters the first 8 characters of the name must be unique
- Formnames must not contain file name extensions; Formname.TST is not a valid form name. SQL*Forms assigns the extensions .INP and .FRM to

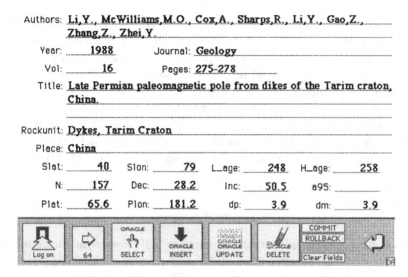

Fig. 20.   Example of the GPMDB using the Macintosh HyperCard. (a) Query using a country name.
(b) The first record retrieved of a total of 64.

distinguish the different form file types. Giving a form a name containing a period could cause SQL*Forms to ignore the characters after the period, possibly resulting in creation errors.

SQL*Forms is an extremely powerful utility that is only mastered in detail after some considerable experience. The custombuilt Form PALEOMAG supplied with the database has evolved with ever-increasing complexity over the two years spent developing the GPMDB.

4. ORACLE for Macintosh with HyperCard

Dr Hidefumi Tanaka of the Tokyo Institute of Technology has experimented with using the GPMDB with ORACLE for Macintosh based on HyperCard using a Mac II. With HyperCard it is possible to design 'Form-like' screens to view the database although SQL*Forms itself cannot be run with ORACLE for Macintosh. Dr Tanaka has kindly provided some examples he has developed from the GPMDB and we are most grateful for his permission to use them here.

Figures 19 and 20 show two examples of queries based around a simple 'Form' that views 20 columns selected from all the different ORACLE tables in the database. The presentation gives the full reference details from the AUTHORS and REFERENCE table, ROCKUNIT and PLACE from the ROCKUNIT table, and basic parameters for the pole position from the PMAGRESULT table. Figure 19(a) shows that the same query procedure using %McElhinny% is used to find all entries for which McElhinny was an author of the paper. Figure 19(b) then gives the first of these entries retrieved with the second window underneath the screen display showing that a total of 40 entries were retrieved. The second example in Figure 20(a) shows a query for all results from China published in years $>=1980$. The first of the 64 results is displayed in Figure 20(b).

Unfortunately it appears that the HyperCard utility is very slow. Dr Tanaka reports that most queries took about $3\frac{1}{2}$ minutes to retrieve using a Mac II. On SQL*Forms with MS-DOS such queries would certainly not take longer than 20 seconds even on 80286, 10 MHz machines. This suggests that ORACLE with HyperCard could be up to an order of magnitude slower than ORACLE on MS-DOS.

Note Added in Proof

ORACLE have announced the development of SQL*Forms for Macintosh and its production release is planned for the second quarter of 1991. SQL*Forms for Macintosh allows users to transfer and run SQL*Forms 3.0 applications created on different platforms to the Macintosh environment. Although the Forms we have developed used SQL*Forms 2.3, it seems likely there will be little difficulty in transporting our version of the Forms to the Macintosh.

V. DATABASE QUERIES WITH SQL*PLUS

1. The ORACLE Database System

The language used to access the database is SQL (Structured Query Language – pronounced SEQUEL). Data are retrieved from the database through queries. SQL is an English-like language that can be used to build queries of substantial complexity and capability. Users with little or no experience can learn SQL's basic features very quickly. The aim of this Section is to familiarise you with these basics and provide experience working with the GPMDB. It is not a comprehensive coverage of all ORACLE'S features, for this refer to the ORACLE documentation (SQL Language Reference Manual, SQL*Plus Reference Manual) and work through the tutorials provided in the 'SQL*Plus User's Guide'.

ORACLE'S kernel takes care of the way in which data are stored in the database. Unlike many other types of databases, you, the user, do not need to know how, where or in what format the data are stored. Through SQL you specify WHAT you want ORACLE to DO, NOT HOW to DO it.

1.1. STARTING ORACLE

After booting up your system, from any directory on the C: disk enter:

C:\ > GORACLE

This starts a batch file that loads ORACLE into extended memory and changes the directory to the GPMDB work directory C:\PALEOMAG where the customised files supplied with the database are located. We therefore follow the convention of assuming that you will be working in this directory. You can of course load ORACLE and work in any sub-directory of the root directory. However, if you then want to use any of the customised files, the full path name of the file must be specified (e.g. C:\PALEOMAG\filename).

If ORACLE has not been started and you attempt to log into any ORACLE utility or tool you will receive the following message:

C:\PALEOMAG > "PME services not available"

1.2. STOPPING ORACLE

When you have finished your database session and/or you intend to shut down your system enter:

C:\PALEOMAG > SHUTORA

This starts a batch file that shuts ORACLE down in an orderly manner and removes it from extended memory.

You should always shut ORACLE down before shutting down your system. In any case, it is good practice to shut ORACLE down at the end of your database session. If you, or another user, attempt to restart ORACLE and it has already been started you will receive the following message:

C:\PALEOMAG > "Error: SQLPME is already loaded"

This is no cause for concern, it merely means that ORACLE is already loaded.

1.3. ORACLE USERID AND PASSWORD

Once ORACLE is started you may log into one of ORACLE'S utilities such as IMP or EXP, or one of ORACLE'S tools such as SQL*Plus, RUNFORM or SQL*Menu. To do this you will need a userID and password. If you do not already have them see your DBA for the issue of a userID and password. These are not case sensitive (i.e. they can be entered as all upper case, all lower case or mixed case). We follow the convention of using all upper case.

A userID and password are required to ensure that only authorised users of the database have access to the data and also, more importantly, to restrict users' privileges. Users may select the data (that is look at it) but may not insert new data, update (change) or delete data. These restrictions preserve the integrity of the GPMDB

2. Introduction to SQL*Plus

2.1. LOGGING INTO AND OUT OF SQL*PLUS

To log into SQL enter:

C:\PALEOMAG> SQLPLUS userID/password

If you do not specify a userID and password SQL*Plus will prompt you for them.

At any log in you may change your password using the GRANT command. Passwords should be changed regularly to maintain data security. In SQL*Plus enter:

SQL> GRANT CONNECT TO youruserID IDENTIFIED BY new-password;

Be careful to record your password and keep it in a secure place. Do not trust your memory! ORACLE encodes passwords in such a way that not even the DBA can decode them.

When you have finished your SQL session enter:

SQL> EXIT

and you will return to MS-DOS.

2.2. ENTERING SQL COMMANDS

There are two kinds of commands that can be entered in SQL*Plus.

1. SQL commands for working with data in the GPMDB. These commands are discussed in this section and in Sections V.3 and V.5. These commands must be terminated with a semi-colon.
2. SQL*Plus commands. Editing and storing of SQL commands is discussed in this section. The editing commands may be terminated by a semi-colon but it is not mandatory. We follow the convention of omitting the semi-colon. Formatting results and setting options are discussed in Section V.4 (SQL*Plus Commands). Many of the other SQL*Plus commands do not need to be terminated with a semi-colon. However, we find it simplest to terminate all SQL* Plus commands, other than editing commands with a semi-colon.

Entering SQL Commands

In general:

- begin entering the command at the SQL> prompt
- pressing [Enter] moves the cursor to the next line, a line number is automatically displayed and more SQL code can be entered
- commands may be entered in upper or lower case and may be entered on one or more lines
- extra blanks inserted to make the command more readable are ignored by SQL*Plus
- a SQL command is always terminated by a semi-colon (;)
- mistakes may be corrected by back spacing and re-entering on the current line. Once [Enter] is pressed previous lines can not be accessed for correction.
- table and view names are case sensitive and must match exactly the name in the database.

In particular, we have followed several convenient conventions:

- SQL commands are entered in upper case
- each clause of the SQL command is entered on a separate line to make the command more readable in the GPMDB

SQL*Plus uses several of the keyboard keys to carry out special functions. Table IX lists the SQL*Plus functions and the corresponding PC keyboard key.

The GPMDB tables and views are listed and described in Appendix 1.

TABLE IX
SQL*Plus function keys

SQL*Plus function name	Keyboard name	Function
[Return]	[Enter]	Terminates a line of input
[Backspace]	[←----]	Moves the cursor to the left
[Pause]	[Pause]	Suspends program operation and display of output
[Resume]	[any key]	Resumes program operation and continues output display after a [Pause]
[Interupt]	[Ctrl][C]	Halts program operation and returns to the SQL> prompt. This is the PANIC button

2.3. SELECTING DATA FROM A TABLE – THE SELECT STATEMENT

Retrieving data from the database is the most common SQL operation. A database retrieval is called a query and to issue a query the SELECT command is used. The basic SELECT command has two parts, called clauses.

    SQL> SELECT some data (column name(s))
    2   FROM a table (table name)

The SELECT clause is always entered first, and is immediately followed by the FROM clause. When you enter the semi-colon ORACLE runs the command.

The database contains a small table called JOURNAL with columns ABBREVI-ATION CHAR(8) and FULLNAME CHAR(60). To view the contents of this table enter:

    SQL> SELECT ABBREVIATION, FULLNAME
    2   FROM JOURNAL;

See Figure 21 for the output from the above query. In this query, we listed every column in the JOURNAL table. If you wish to select all the columns of a table use an asterisk (*) in place of the list of column names. This is slower than specifying each column by name. Enter:

    SQL> SELECT *
    2   FROM JOURNAL;

As you have seen (Figure 21) the result of any query is itself a table made up of columns and rows. You list only those columns you want to view. The order in which you list the columns in the SELECT statement controls the left-to-right sequence in the result table. When you specify SELECT *, the column sequence of the query result is determined by the order in which the columns were specified when the table was created. To reverse the order of the columns in JOURNAL enter:

```
SQL> SELECT ABBREVIATION, FULLNAME
  2  FROM JOURNAL;

ABBREVIA FULLNAME
-------- -----------------------------------------------------------
JGR      J.Geophys.Res.
GJRAS    Geophys.J.Roy.Astron.Soc.
GJ       Geophysical Journal
EPSL     Earth Planet.Sci.Letters
GSAB     Geol.Soc.Amer.Bull.
GEO      Geology
PEPI     Phys.Earth Planet.Interiors
JG       J.Geophys.
TECT     Tectonics
TPHYS    Tectonophysics
PR       Precambrian Res.
JGG      J.Geomag.Geoelect.
CJES     Canad.J.Earth Sci.
OTHER    Type in the Name
NAT      Nature
PAG      Pure & Appl.Geophys.
PPP      Palaeogeog.Palaeoclim.Palaeoecol.
NZJGG    N.Z.J.Geol.Geophys.
JGSA     J.Geol.Soc.Austr.
JGD      J.Geodynamics
GRL      Geophys.Res.Lett.
JGSL     J.Geol.Soc.London
JGEO     J.Geol.
GM       Geol.Mag.
TAGU     Trans.Amer.Geophys.Union
AJES     Aust.J.Earth Sci.
AJS      Amer.J.Sci.
CRASP    C.R.Acad.Sci.Paris
BGSF     Bull.Geol.Soc.Finland
GFF      Geol.Foren.Stockholm Forh.
GSCP     Geol.Surv.Canada Paper
GSFB     Geol.Surv.Finland Bull.
ZG       Zeits.Geophys.
GJI      Geophys.J.Int.
GSCB     Geol.Surv.Canada Bull.

35 records selected.
```

Fig. 21.   The SELECT Statement.

SQL> SELECT FULLNAME, ABBREVIATION
  2   FROM JOURNAL;

2.4. SELECTING SPECIFIC ROWS – THE *WHERE* CLAUSE

To retrieve specific rows from a table a conditional clause, the WHERE clause, must be added to the SELECT statement. A WHERE clause causes ORACLE to search the data in the table and retrieve only those rows that meet the search-condition. The condition is defined by a logical expression that is true or false. To display the standard entry in JOURNAL for the abbreviation JGR (Figure 22), enter:

```
SQL> SELECT FULLNAME
  2  FROM JOURNAL
  3  WHERE ABBREVIATION = 'JGR';

FULLNAME
---------------------------------------------------------------
J.Geophys.Res.
```

Fig. 22.   Selecting specific rows from a table – the WHERE clause and the = operator.

SQL> SELECT FULLNAME
 2   FROM JOURNAL
 3   WHERE ABBREVIATION = 'JGR';

If you make a mistake and entered the following:

SQL> SELECT FULLNAME
 2   FROM JOURNAL
 3   WHERE ABBREVIATION = JGR;

an ORACLE error message will appear:

SQL> WHERE ABBREVIATION = JGR
                          *

"ERROR at line 3: ORA-0704: invalid column name"

This message is attempting to tell you that there is something wrong with the abbreviation you have specified. Or if you entered:

SQL> SELECT FULLNAME
 2   FROM JOURNAL
 3   WHERE ABBREVIATION = 'jgr';

Then ORACLE would give the message: "no records selected" This message occurs because no match for the abbreviation was found in the database.

Note:

1. ABBREVIATION does not have to be listed in the SELECT statement. SELECT only the column(s) that you wish to see.
2. Characters must be surrounded by single quotes.
3. The case in the query must match that in the database.

The other comparison operators are listed in Table X.

Suppose you wish to know the correct abbreviation for the journal Tectonics and you are not sure what the full name of the journal will be in the database. This can be achieved by the use of LIKE operator instead of = in the WHERE clause and % to represent any number of unknown characters, including none (Figure 23).

TABLE X
Logical operators

=	Equal to
!=, ^=, <>	Not equal to
<	Less than
<=	Less than or equal to
>	Greater than
>=	Greater than or equal to
AND	Combines logical expressions, and is true if all expressions are true
OR	Combines logical expressions, and is true if any expression is true
BETWEEN value1 AND value2	Any values between the two values inclusively, where value1 < value2
IN (v1, v2, v3, . . . , vn)	Any value in the specified list.
LIKE 'text string'	Any value which match the text string
LIKE 'te_xt string'	_ represents a single unknown Alphanumeric in this position
LIKE '%string'	% indicates any number of character positions containing any alphanumeric
NOT	Reverses the result of the logical expressions BETWEEN, IN, LIKE, IS NULL
IS NULL	Will test for the existence of null's

NOTE: The square brackets [ ] indicate an optional item NULL is not the same as zero or a blank and may not be equated to them.
NULL != zero, NULL != blank and NULL != NULL

```
SQL> SELECT *
  2  FROM JOURNAL
  3  WHERE FULLNAME LIKE 'Tect%';

ABBREVIA FULLNAME
-------- ----------------------------------------------------------------
TECT     Tectonics
TPHYS    Tectonophysics
```

Fig. 23. Selecting specific rows from a table – the WHERE clause and the LIKE 'string%' operator.

SQL> SELECT *
  2   FROM JOURNAL
  3   WHERE FULLNAME LIKE 'Tect%';

To list the journals with Australia or some abbreviation, or variation of Australia in the journal name (Figure 24), enter:

SQL> SELECT *
  2   FROM JOURNAL
  3   WHERE FULLNAME LIKE '%Aust%';

```
SQL> SELECT *
  2  FROM JOURNAL
  3  WHERE FULLNAME LIKE '%Aust%';

ABBREVIA FULLNAME
-------- -----------------------------------------------------------
JGSA     J.Geol.Soc.Austr.
AJES     Aust.J.Earth Sci.
```

Fig. 24. Selecting specific data from a table – the WHERE clause and the LIKE '%string% operator.

2.5. THE SQL BUFFER

When a SQL command is entered, it is stored in a part of the memory called the SQL buffer. It remains there until a new SQL command is entered or the buffer is cleared. This command is the current SQL command. If you want to edit or re-run the current command (using SQL*Plus commands), you may do so without re-entering it. The semi-colon, which must be entered to indicate the end of a SQL command, is not stored in the buffer as part of the command.

2.6. THE SQL*PLUS LINE EDITOR

The SQL*Plus line editor allows changes to be made to a single line of code in the SQL buffer. A line must be made the current line using the LIST command followed by the line number before changes can be made. The current line is indicated by an asterisk (*) after the line number. When the contents of the SQL buffer are listed the current line becomes the last line.

Table XI lists the SQL*Plus commands that allow you to examine, change, or re-run a SQL command without re-entering it.

To list the current SQL command in the SQL buffer enter:

TABLE XI
SQL*Plus line editor commands

Command	Abbreviation	Purpose
APPEND text	A text	Add text at the end of the current line
CHANGE	C/old/new/	Change old text to new on the current line
CHANGE/text/	C/text/	Delete text from the current line
CLEAR BUFFER	CL BUFF	Delete all lines from the buffer
DEL	(none)	Deletes the current line
INPUT	I	Add an indefinite number of lines after the current line
INPUT text	I text	Adds a single line of text after the current line
LIST	L	List all lines in the current buffer
LIST *	L*	List the current line in the buffer
LIST n	Ln	List the nth line in the current buffer
LIST m n	Lm n	List lines m to n in the current buffer
LIST LAST	L LAST	List the last line in the current buffer
RUN	R	Re-run the current SQL command, echoes the command
/	/	Re-run the current SQL command, does not echo the command

SQL> L

This will list the current contents of the SQL buffer.

SQL> SELECT FULLNAME
2   FROM JOURNAL
3* WHERE ABBREVIATION = '%Aust%'

To change Aust to Geol, for example, enter:

SQL> L3

This will list line 3 of the SQL buffer

3* WHERE ABBREVIATION = '%Aust%'

Use the CHANGE (C) command to change Aust (old string) to Geol (new String).

SQL> C/Aust/Geol/

To list the current contents of the SQL buffer enter:
SQL> L

Then the following will display:

SQL> SELECT FULLNAME
2   FROM JOURNAL
3* WHERE ABBREVIATION = '%Geol%',

If you wish to run the new command without echoing the SQL command enter:

SQL> /

2.7. ORDERING THE ROWS FROM A QUERY − THE *ORDER BY* CLAUSE

In the examples so far, the rows resulting from a query have been displayed in an order determined by ORACLE. The same query need not fetch these rows in the same order the next time it is issued. You can control the order in which the selected rows are displayed by adding an ORDER BY clause to the end of your SELECT command. The ORDER BY must be the last clause in the SELECT statement. The order can be either descending or ascending and the latter is the default. Suppose you want to list the JOURNAL table so that the abbreviations are in alphabetical order (Figure 25) then:

SQL> SELECT *
2   FROM JOURNAL
3   ORDER BY ABBREVIATION;

```
SQL> SELECT *
   2  FROM JOURNAL
   3  ORDER BY ABBREVIATION;

ABBREVIA FULLNAME
-------- --------------------------------------------------------
AJES     Aust.J.Earth Sci.
AJS      Amer.J.Sci.
BGSF     Bull.Geol.Soc.Finland
CJES     Canad.J.Earth Sci.
CRASP    C.R.Acad.Sci.Paris
EPSL     Earth Planet.Sci.Letters
GEO      Geology
GFF      Geol.Foren.Stockholm Forh.
GJ       Geophysical Journal
GJI      Geophys.J.Int.
GJRAS    Geophys.J.Roy.Astron.Soc.
GM       Geol.Mag.
GRL      Geophys.Res.Lett.
GSAB     Geol.Soc.Amer.Bull.
GSCB     Geol.Surv.Canada Bull.
GSCP     Geol.Surv.Canada Paper
GSFB     Geol.Surv.Finland Bull.
JG       J.Geophys.
JGD      J.Geodynamics
JGEO     J.Geol.
JGG      J.Geomag.Geoelect.
JGR      J.Geophys.Res.
JGSA     J.Geol.Soc.Austr.
JGSL     J.Geol.Soc.London
NAT      Nature
NZJGG    N.Z.J.Geol.Geophys.
OTHER    Type in the Name
PAG      Pure & Appl.Geophys.
PEPI     Phys.Earth Planet.Interiors
PPP      Palaeogeog.Palaeoclim.Palaeoecol.
PR       Precambrian Res.
TAGU     Trans.Amer.Geophys.Union
TECT     Tectonics
TPHYS    Tectonophysics
ZG       Zeits.Geophys.

35 records selected.
```

Fig. 25.   Ordering the rows of a query result – the ORDER BY clause.

2.8. SELECTING UNIQUE ROWS – THE *DISTINCT* OPERATOR

Unless you indicate otherwise SQL*Plus displays the results of a query without eliminating duplicate entries. Table ROCKUNIT has a CONTINENT and also a TERRANE column (see Appendix 1 for a description of the table). Suppose you want a list of all the continents entered in the database, but with no duplicate entries (Figure 26(a)), enter:

SQL> SELECT DISTINCT CONTINENT
   2   FROM ROCKUNIT;

If you wish to see all the distinct terranes in the database (Figure 26(b)) enter:

```
(a)
SQL> SELECT DISTINCT CONTINENT
  2  FROM ROCKUNIT;

CONTINENT
---------------
Africa
Antarctica
Arctic Ocean
Asia
Atlantic Ocean
Australia
Europe
Greenland
Indian Ocean
North America
Pacific Ocean
South America

12 records selected.

(b)
SQL> SELECT DISTINCT TERRANE
  2  FROM ROCKUNIT;

TERRANE
------------------------------
Alexander
Antarctic Peninsula
Arabia
Avalon
Bohemian Massif
Calabria
Chortis
Churchill Province
       .
       .
       .
Yilgarn

59 records selected.
```

Fig. 26.   Selecting unique rows from a table – the DISTINCT operator.

SQL> SELECT DISTINCT TERRANE
  2   FROM ROCKUNIT;

If you wish to see the distinct terranes continent by continent then the column name with the DISTINCT clause must come first in the SELECT. To list all the terranes for one continent together then an ORDER BY clause is needed. To exclude continents for which no terrane has been listed then the NOT NULL condition must be used on TERRANE (Figure 27).

SQL> SELECT DISTINCT TERRANE, CONTINENT
  2   FROM ROCKUNIT
  3   WHERE TERRANE IS NOT NULL
  4   ORDER BY CONTINENT, TERRANE;

```
SQL> SELECT DISTINCT TERRANE, CONTINENT
  2 FROM ROCKUNIT
  3 WHERE TERRANE IS NOT NULL
  4 ORDER BY CONTINENT, TERRANE;
```

TERRANE	CONTINENT
Kaapvaal	Africa
Limpopo Belt	Africa
Madagascar	Africa
West Africa	Africa
Antarctic Peninsula	Antarctica
East Antarctica	Antarctica
Ellsworth Mountains	Antarctica
Thurston Island	Antarctica
Spitsbergen	Arctic Ocean
Arabia	Asia
India	Asia
Indosinian	Asia
Iran	Asia
Kunlun	Asia
Lhasa	Asia
Mino	Asia
North China	Asia
Qiangtang	Asia
Siberia	Asia
South China	Asia
Tarim	Asia
Gawler	Australia
Hammersley Basin	Australia
Kimberley	Australia
Mt.Isa	Australia
Musgrave	Australia
Northampton	Australia
Pilbara	Australia
Yilgarn	Australia
Bohemian Massif	Europe
Calabria	Europe
Corsica	Europe
Fennoscandian Shield	Europe
Iberia	Europe
.	
.	
.	
Mirihiku	Pacific Ocean

59 records selected.

Fig. 27.   Selection of distinct terranes, continent by continent.

NOTE: When ORDER BY and DISTINCT are both specified, the ORDER BY clause must refer only to columns listed in the SELECT clause.

2.9. VIEWS

A VIEW is a window through which you can look at information in the database tables. Although views contain no data of their own they look like tables and, with some restrictions, can be treated as such. Views occupy no storage space, so they are sometimes called 'virtual tables'. Instead of using a complex query you can define a view that allows you to get the same information with a simple query. This will also be faster as the SQL statement has already been parsed when the view was created. You may use any valid query in a CREATE VIEW command except an ORDER BY clause. If you want to order the rows, you must do it when you query the view. Suppose you wish to create a view using our last SQL statement:

```
SQL> CREATE VIEW TLIST AS
  2    SELECT DISTINCT TERRANE, CONTINENT
  3    FROM ROCKUNIT
  4    WHERE TERRANE IS NOT NULL;
```

To query this view and order the results, enter:

```
SQL> SELECT *
  2    FROM TLIST
  3    ORDER BY CONTINENT, TERRANE;
```

You may define a view of a view. See also Section V.6.3 (Creating and Dropping Views).

Some views are supplied with the GPMDB and are listed and described in Appendix 1. The code used to create the views is given in Appendix 3.

3. Developing More Complex SQL Commands

One of the most likely things a new user of the GPMDB would want to do is to see which of their publications has been referenced in the database. Step-by-step we will develop a more complex SQL command to select the required reference data from the database. This is the method you should adopt if you develop your own SQL commands. You will first need to know which table or tables store the data that are required. This can be determined by using the data dictionary.

3.1. THE DATA DICTIONARY

ORACLE automatically maintains a group of tables and views called the data dictionary. These contain current information about the database. Data dictionary tables can be read with standard SQL queries. Since the data dictionary tables are themselves described in the data dictionary, the dictionary may be queried to determine the names of its own tables, views, columns etc.

```
SQL> SELECT *
  2    FROM DTAB;
```

```
TNAME             REMARKS
---------------   -------------------------------------------------------
Reference Date    ORACLE catalog as of 10-Oct-85, installed on
                  30-OCT-88 00:38:23.
CATALOG           Tables and views accessible to user (excluding
                  data dictionary)
COL               Specifications of columns in tables created by the
                  user
COLUMNS           Columns in tables accessible to user (excluding
                  data dictionary)
DTAB              Description of tables and views in Oracle Data
                  Dictionary
EXTENTS           Data structure of extents within tables
INDEXES           Indexes created by user and indexes on tables
                  created by user
PRIVATESYN        Private synonyms created by the user
PUBLICSYN         Public synonyms
SPACES            Selection of space definitions for creating tables
                  and clusters
STORAGE           Data and Index storage allocation for user's own
                  tables
SYNONYMS          Synonyms, private and public
SYSCATALOG        Profile of tables and views accessible to the user
SYSCOLUMNS        Specifications of columns in accessible tables and
                  views
SYSPROGS          List of programs precompiled by user
SYSTABAUTH        Directory of access authorization granted by or to
                  the user
SYSUSERLIST       List of Oracle users
SYSVIEWS          List of accessible views
TAB               List of tables, views, clusters and synonyms
                  created by the user
TABALLOC          Data and index space allocations for all user's
                  tables
TABQUOTAS         Table allocation (space) parameters for tables
                  created by user
VIEWS             Defining SQL statements for views created by the
                  user
```

Fig. 28. A selection of useful tables and views listed from the data dictionary table, DTAB. The entries will appear on one line on your screen.

Figure 28 lists the more useful tables and views in the data dictionary table DTAB. DTAB has two columns:

TNAME   – contains the name of each table
REMARKS – briefly describes the table.

CATALOG is listed as containing the tables and views accessible to the user. To look at the contents of CATALOG it is helpful to know the TABLE DEFINITION of CATALOG.

3.2. DETERMINING TABLE DEFINITIONS – THE *DESCRIBE* COMMAND

The DESCRIBE command displays a brief description of a table or view. The name for each column is listed, and the following information given:

```
(a)
SQL> DESCRIBE CATALOG;

Name                                    Null?      Type
--------------------------------------- --------   ----
TNAME                                   NOT NULL   CHAR(30)
CREATOR                                      '     CHAR(30)
TABLETYPE                               NOT NULL   CHAR(7)
CLUSTERID                                          NUMBER
LOGBLK                                             NUMBER
REQBLK                                             NUMBER
IXCOMP                                             CHAR(10)
REMARKS                                            CHAR(240)


(b)
SQL> SELECT TNAME, TABLETYPE
   2  FROM CATALOG;

TNAME                                   TABLETY
--------------------------------------- -------
ALLCAGE                                 VIEW
ALTRESULT                               TABLE
AUTHORS                                 TABLE
CONTINENTS                              TABLE
CROSSREF                                TABLE
FIELDTESTS                              TABLE
INFORMATION                             TABLE
JOURNAL                                 TABLE
KEYS                                    TABLE
PMAGRESULT                              TABLE
PRICAGE                                 VIEW
REFERENCE                               TABLE
REMARKS                                 TABLE
ROCKUNIT                                TABLE
RSLTALL                                 VIEW
RSLTPRI                                 VIEW
TIMESCALE                               TABLE

17 records selected.
```

Fig. 29. Determining table definitions and types using the System table CATALOG. (a) The DE-
SCRIBE operator. (b) Listing the GPMDB tables and views.

● NOT NULL indicates that there must always be an entry in the column. If it
  is not specified there may be a null entry in the column.
● the datatype of the column, NUMBER or CHAR
● the size of the column

The table definition of CATALOG can be seen by issuing the SQL command
DESCRIBE (Figure 29(a)):

SQL> DESCRIBE CATALOG;

The information we need is contained in the TNAME and TABLETYPE columns
of CATALOG (Figure 29(b)).

SQL> SELECT TNAME, TABLETYPE
   2   FROM CATALOG;

```
(a)
SQL> DESCRIBE AUTHORS;

Name                                Null?       Type
---------------------------------   --------    ----
REFNO                               NOT NULL    NUMBER(4)
AUTHORS                             NOT NULL    CHAR(240)


(b)
SQL DESCRIBE REFERENCE;

Name                                Null?       Type
---------------------------------   --------    ----
REFNO                               NOT NULL    NUMBER(4)
YEAR                                NOT NULL    NUMBER(4)
JOURNAL                             NOT NULL    CHAR(60)
VOLUME                                          NUMBER(4)
PAGES                               NOT NULL    CHAR(11)
TITLE                               NOT NULL    CHAR(200)
```

Fig. 30.   Examples of table definitions. (a) Table AUTHORS, (b) Table REFERENCE.

To produce a standard reference, the data required will come from both the REFERENCE and the AUTHORS table. The structure of these tables can be seen by using the DESCRIBE command (Figure 30).

SQL> DESCRIBE AUTHORS;
SQL> DESCRIBE REFERENCE;

3.3. SELECTING DATA FROM MORE THAN ONE TABLE – TABLE JOINS

For every REFERENCE there is always at least one corresponding author in the AUTHORS table. Notice that the column REFNO appears in both tables. This is a primary key which can be used to JOIN the two tables together so that they can be treated as if they were one table. As the join is made by equating two columns in different tables, it is called an EQUI-JOIN. This is the most commonly used type of join.

We do *NOT* recommend that you enter the following SQL commands flagged by a '#' as they will retrieve all the references in the database (more than 2000).

The simplest SQL statement that joins the AUTHORS and REFERENCE tables is:

SQL> SELECT *
   2   FROM AUTHORS, REFERENCE
   3   WHERE AUTHORS.REFNO = REFERENCE.REFNO;

When two different tables contain the same column name, the table name must be used in conjunction with the column name so that SQL can find the correct column. SQL*Plus will give an error if this is not done.

3.4. ABBREVIATING A TABLE NAME – TABLE ALIASES

The amount of typing that needs to be done can be reduced by setting up aliases for the table names. This is particularly useful if many tables are being joined. The alias immediately follows the table name in the FROM clause.

```
# SQL> SELECT *
    2    FROM AUTHORS A, REFERENCE R
    3    WHERE A.REFNO = R.REFNO;
```

The above SELECT statements will select all the columns from both tables. It is not necessary to list the REFNO columns in the output, so specify the columns required, and the order in which they should be listed.

```
# SQL> SELECT AUTHORS, YEAR, JOURNAL,
       VOLUME, PAGES, TITLE
    2    FROM AUTHORS A, REFERENCE R
    3    WHERE A.REFNO = R.REFNO;
```

This statement selects all the entries in both tables, but suppose only some particular entries are required. Then a further selection criterion is needed.

3.5. USING MORE THAN ONE SELECTION CRITERION – THE *AND* OPERATOR

Suppose a list of references is required for author McElhinny. The rows returned by the query can be restricted further by adding another condition to the SQL statement using the AND operator.

```
# SQL> SELECT AUTHORS, YEAR, JOURNAL,
       VOLUME, PAGES, TITLE
    2    FROM AUTHORS A, REFERENCE R
    3    WHERE A.REFNO = R.REFNO
    4    AND AUTHORS = 'McElhinny';
```

"no records selected"

No rows are returned by this query as the condition has been made too restrictive. No initials have been included nor has any allowance been made for the fact there may be other authors on the paper or that McElhinny may not be the first author. To make the condition less restrictive use the LIKE operator and % to represent the unknown characters. Try entering the following command (Figure 31):

```
SQL> SELECT AUTHORS, YEAR, JOURNAL,
     VOLUME, PAGES, TITLE
  2    FROM AUTHORS A, REFERENCE R
  3    WHERE A.REFNO = R.REFNO
  4    AND AUTHORS LIKE '%McElhinny%';
```

```
AUTHORS
------------------------------------------------------------------
YEAR JOURNAL                                    VOL PAGES
-----  ------------------------------------------- ----- ----------
TITLE
------------------------------------------------------------------

McElhinny,M.W., Senanayake,W.E.
1980 J.Geophys.Res.                             85 3523-3528
Paleomagnetic evidence for the existence of the geomagnetic field
3.5 Ga ago

McFadden,P.L., Ma,X.H., McElhinny,M.W., Zhang,Z.K.
1988 Earth Planet.Sci.Letters                  87 152-160
Permo-Triassic magnetostratigraphy in China: northern Tarim

McWilliams,M.O., McElhinny,M.W.
1980 J.Geol.                                    88 1-26
Late Precambrian paleomagnetism in Australia: the Adelaide
Geosyncline

McElhinny,M.W., Embleton,B.J.J., Ma,X.H., Zhang,Z.K.
1981 Nature                                     293 212-216
Fragmentation of Asia in the Permian

Wellman,P., McElhinny,M.W., McDougall,I.
1969 Geophys.J.Roy.Astron.Soc.                  18 371-395
On the polar wander path for Australia during the Cenozoic

McElhinny,M.W., Luck,G.R.
1970 Geophys.J.Roy.Astron.Soc.                  20 191-205
The palaeomagnetism of the Antrim Plateau Volcanics of Northern
Australia
       .
       .
       .
McElhinny,M.W.
1964 Geophys.J.Roy.Astron.Soc.                  8 338-340
Statistical significance of the fold test in palaeomagnetism


34 records selected.
```

Fig. 31. Reference list for a selected author. The output shown here has been formatted for clarity. When these references are listed on your screen they will be unformatted and occupy 80 characters across the screen.

The modifications to the SQL command that follow can either be made using the SQL*Plus line editor commands or another editor. To list the references in order by year add an ORDER BY clause to the end SQL command:

```
SQL> SELECT AUTHORS, YEAR, JOURNAL,
     VOLUME, PAGES, TITLE
  2  FROM AUTHORS A, REFERENCE R
  3  WHERE A.REFNO = R.REFNO
  4  AND AUTHORS LIKE '%McElhinny%'
  5  ORDER BY YEAR;
```

```
AUTHORS
-----------------------------------------------------------------
YEAR JOURNAL                                         VOL PAGES
-----  ----------------------------------------- -----  ----------
TITLE
-----------------------------------------------------------------


McElhinny,M.W., Gough,D.I.
1963 Geophys.J.Roy.Astron.Soc.                        7 287-303
The palaeomagnetism of the Great Dyke of Southern Rhodesia

McElhinny,M.W.
1964 Geophys.J.Roy.Astron.Soc.                        8 338-340
Statistical significance of the fold test in palaeomagnetism

McElhinny,M.W., Opdyke,N.D.
1964 J.Geophys.Res.                                  69 2465-2475
The paleomagnetism of the Precambrian dolerites of eastern
Southern Rhodesia, an example of geologic correlation by rock
magnetism

McElhinny,M.W., Jones,D.L.
1965 Nature                                         206 921-922
Palaeomagnetic measurements on some Karroo dolerites from
Rhodesia

McElhinny,M.W.
1966 Earth Planet.Sci.Letters                         1 439-442
Rb-Sr and K-Ar measurements on the Modipe gabbro of Bechuanaland
and South Africa

McElhinny,M.W.
1966 Geophys.J.Roy.Astron.Soc.                       10 375-381
The palaeomagnetism of the Umkondo Lavas, eastern Southern
Rhodesia
                 .
                 .
McFadden,P.L., Ma,X.H., McElhinny,M.W., Zhang,Z.K.
1988 Earth Planet.Sci.Letters                        87 152-160
Permo-Triassic magnetostratigraphy in China: northern Tarim


34 records selected.
```

Fig. 32. Ordered reference list for a selected author. See Figure 31 for a comment on print layout.

To achieve the ultimate in ordering modify L5 to read:

5* ORDER BY YEAR, JOURNAL, VOLUME, PAGES

This will order first by YEAR, within each year by JOURNAL, within a journal by VOLUME and within a volume by PAGES (Figure 32).

3.6. SUBSTITUTION VARIABLES – &VARIABLENAME

The author name can be entered interactively. This is done by using a substitution variable prefixed by & (e.g. &AUTHORNAME) instead of the actual name. Modify L4 to read:

4* AND AUTHORS LIKE '%&AUTHORNAME%'

SQL> SELECT AUTHORS, YEAR, JOURNAL, VOLUME, PAGES, TITLE
2   FROM AUTHORS A, REFERENCE R
3   WHERE A.REFNO = R.REFNO
4   AND AUTHORS LIKE '%&AUTHORNAME%'
5   ORDER BY YEAR, JOURNAL, VOLUME, PAGES;

When you run this command SQL will prompt you for the author name:

"Enter value for authorname:"

If you enter McElhinny as the author the query output listed on the screen will be similar to Figure 32.

The required SELECT statement has now been developed, however the format of the output to the screen is not very readable. With some thought, output fairly close to that of standard reference, can be produced. This will be considered further in Section V.4 (SQL*Plus Commands).

3.7. WARNING – BEWARE OF TRAPS FOR THE UNWARY

80 characters will fit across the screen. Several of the database tables have more columns than will fit across the screen on one line, so output wraps around becoming hard to read. For example the table PMAGRESULT has more than 40 columns. It is impossible to SELECT all of these columns and display them on the screen except by querying using SQL*Forms.

DO NOT ENTER the following:

SQL> SELECT *
2*   FROM PMAGRESULT;

This can produce a flood of unreadable output which streams up the screen at a rate likely to induce panic! First let's look at why this seemingly benign SQL command can run amuck. There are two separate problems:

1. As PMAGRESULT has more than 40 columns it takes the entire screen to wrap one row. You must select the columns you really wish to view and format them to fit across the screen in one line. This is the procedure for any table query.
2. PMAGRESULT contains more than 4000 rows and the above query will retrieve all of them. Either restrict the number of rows that will be returned by a query by using search condition and/or instruct SQL*Plus to limit the number of rows displayed on the screen at any one time.

How do we cut this run away process off at the pass? We press the *PANIC* button which on most MS-DOS machines is [Ctrl][Break]. Note there may be a slight delay before the operation underway actually ceases.

You can see the need to format the results of your queries to the screen or printer and control the way in which they are displayed on the screen or printed. SQL*Plus commands allow you to achieve this and are described in the next section.

4. SQL*Plus Commands

4.1. ENTERING SQL*PLUS COMMANDS

SQL*Plus commands:

- can be entered on one line, when the end of the screen line is reached the cursor automatically moves to the next screen line, ready to continue entering code; do not press [Enter] until the command is complete.
- can be continued over more than one line by entering a hyphen (-) at the end of every line but the last one.
- need not be terminated with a semi-colon (this is optional): we have followed the convention of using a semi-colon to terminate all commands except the SQL*Plus Line Editor commands.
- can be corrected if an error is made entering the command:
  - before [Enter] is pressed, backspace and re-enter the correct code
  - after [Enter] has been pressed, re-enter the complete command correctly

4.2. SQL*PLUS COMMAND SUMMARY

SQL*Plus commands control the manner in which ORACLE processes SQL commands. SQL*Plus commands allow:

1. editing commands in SQL and other buffers – SQL*Plus line editor commands are covered in Section V.2.6
2. storing and retrieval of SQL and SQL*Plus commands
3. control of the way data from queries is displayed – formatting output from queries
4. setting and displaying the ORACLE system environment

Tables XII and XIII list the commands we have found useful. There are many more. Refer to the 'SQL*Plus Reference Manual' and 'SQL*Plus Users Guide' for a full listing of these commands and examples of how to use them.

Table XIII lists some of the environment variables that ORACLE uses. These variables are assigned using the SET command and may be used in one of the following:

1. The LOGIN.SQL file in the database work directory (C:\PALEOMAG).

TABLE XII

A summary of SQL*Plus commands

Command	Description
(a) Storing and retrieval of SQL*Plus commands	
SAVE filename	Saves the current SQL*Plus commands and/or SQL command in filename.SQL
START filename	Executes the SQL*Plus commands and/or the SQL command in filename.SQL
SET BUFFER name	Sets the current buffer to buffer name. The default buffer is the SQL buffer
GET filename	Puts the file into the current buffer which may be either into the SQL buffer or into the buffer named in the SET BUFFER command
SPOOL filename	Spools displayed output to an MS-DOS file filename.LST or .LIS
SPOOL OFF	Stops spooling to a file
(b) Formatting the output from queries	
BREAK ON column	Supresses the display of duplicate values for a column.
SKIP n	Skip n lines
TTITLE 'string'	Title placed at the top of each page
BTITLE 'string'	Caption placed at the bottom of each page
COLUMN column	Specifies how a column and the column heading are formatted. Some options are:
	JUSTIFY LEFT/CENTER/RIGHT
	FORMAT
	PRINT/NOPRINT
	HEADING 'string|string'
	SKIP n
	TRUNC/WRAP/WORD_WRAP

Each of the above commands may be SET ON (the default) or OFF in the current SQL*Plus session.

COLUMN column	The current definition for the column is displayed when no options are specified
CLEAR command	Clears the current (SQL or other) buffer, clears BREAKS, clears T or BTITLE, all COLUMN definitions and SCREEN
SHOW command	Displays the current value of B and TTITLE and SPOOL
REMARK	Begins a remark in a command file. SQL*Plus does not interpret the remark as a command.
HOST	Executes a host operating system (MS-DOS) command without leaving SQL*Plus.

2. In any SQL*Plus session and remain current during that session, unless SET OFF.

3. Stored in a command file to be SET each time the command file is STARTed.

If no formatting is specified data will be displayed on the screen (or printed) according to the parameters set in the LOGIN.SQL file that is consulted when you log into SQL*plus. The default page is set up for the screen and is 80 characters across and 25 lines down. Column headings are repeated at the top of each screen page of output. Unless PAUSE ON is SET output will scroll continu-

TABLE XIII
SET COMMANDS – setting and displaying ORACLE system environment

Command	Description
(a) Setting up the screen or print page	
SET LINESIZE nn	. Sets the number of characters displayed on a line
	. 80 characters is the maximum that will fit across the screen or printed page
SET PAGESIZE nn	. Sets the number of lines per page
	. For the screen set to 25
	. For 11" paper a value of 54 (and NEWPAGE value 6) leaves 1" margins at the top and bottom of the page
SET NEWPAGE nn	. Sets the number of blank lines to be printed between the bottom title of each page and the top title of the next page
(b) Controlling output to the screen or printer	
SET ECHO ON/OFF	. Controls whether the START command displays each command in a file as the command is executed.
SET FEEDBACK ON/OFF	. Displays the number of records returned by a query.
SET PAUSE ON/OFF	. Allows you to control the scrolling of your terminal when output from a query is being displayed. Output will pause every 25 lines. A message can be printed each time SQL*Plus pauses.
SET TERMOUT ON/OFF	. Controls the display of output generated by commands executed from a file. OFF suppresses the display so that you can spool to a print file without seeing the output on the screen. BEWARE do not use with substitution variables as you will not be prompted for the value of the variable.
SET VERIFY ON/OFF	. Controls whether SQL*Plus displays the text of a command line before and after replacement of the substitution variables with values.
SET SPACE 0-10	. Sets the space between each of the columns of output.
SET NUMWIDTH n	. Sets the default width for displaying numbers. The LOGIN.SQL file sets NUMWIDTH 7.

ously up the screen making it very hard to read. A column defined as 'character' defaults to the width assigned when the table was created. If a new column width is specified using FORMAT An (n is any integer up to 240) in the COLUMN command and this width is shorter than the column name, SQL*Plus will truncate the name in the column heading. SQL*Plus left justifies character fields. If a heading or column entry is wider than the column specification then it will WRAP to the next line of output. WORD_WRAP can be set so that the output wraps on whole words. TRUNC truncates output. Columns defined as numeric default to the value of NUMWIDTH. It is SET in the LOGIN.SQL file as seven. It can be reset during any SQL*Plus session or over-ridden by the FORMAT option of the COLUMN command. SQL*Plus right justifies numbers. The column width will be the width of the column heading or the width specified by FORMAT plus one for the sign, whichever is greater. SQL*Plus never truncates a numeric column heading. It may be shortened by using the HEADING 'newname' option of the

COLUMN command. If a numeric value overflows the column width a row of asterisks (*) will be displayed to alert you.

Tables XII and XIII summarise the extensive and powerful set of commands to format output and control either what appears on the screen or what is spooled to a print file. The best way to understand these commands is to see how to use them.

4.3. SQL*PLUS COMMAND FILES

A command file consists of SQL*Plus commands and a SQL command (a SELECT statement) that may contain substitution variables. SQL*Plus assigns the default extension .SQL to the file. We have created a number of customised .SQL command files that are supplied with the GPMDB. One of these, SREFLIST.SQL, is a command file that produces a formatted list of publications for a user nominated author. The SQL command that selects the data is the same as the one we developed in Section V.3 (Developing More Complex SQL Commands). Other .SQL command files are discussed below in Section V.4.3 (SQL*Plus Command Files) and are listed in Table XIV.

The SET BUFFER and GET Commands

The GET command loads an MS-DOS file into the current buffer. If a file contains only a SQL command (a SELECT statement) it can be loaded into the SQL buffer. Initially, the SQL buffer is the current buffer and it cannot store SQL* Plus commands (such as those listed in Tables XII and XIII). Command files containing SQL*Plus commands and possibly a SELECT statement must be loaded into another buffer assigned by the SET BUFFER command. The SET BUFFER command sets the name entered to the current buffer. All the SQL*Plus line editor commands will operate when the named buffer is the current buffer.

The following commands can be used to view the code in SREFLIST.SQL. To set up a named buffer and GET the file, enter:

SQL> SET BUFFER SCRATCH   set up a buffer called SCRATCH

SQL> GET SREFLIST   GET the contents of an MS-DOS file called SREFLIST.SQL and display it on the screen

Figure 33 lists the contents of the file SREFLIST. This gives examples of how the SQL*Plus commands are used. The operation of each of the commands in Figure 33 is listed below.

Lines 1–3

REM begins a REMARK in a command file. SQL*Plus does not interpret the remark as a command to be executed.

TABLE XIV
Customised .SQL command files

Screen	Printer	Comment
SCREEN		Configures VDU screen
	PRINTER	Configures output for the printer
SALLCAGE	PALLCAGE	Lists a summary of paleomagnetic results for a user supplied continent and time window – results for both primary and secondary magnetization are selected
SPRICAGE	PPRICAGE	Lists as above except only results for primary magnetizations are selected
SRSTALL	PRSLTALL	Lists the same results as SALLCAGE and also authors, year of reference, number of sites and number of samples
SRSLTPRI	PRSLTPRI	Lists the same results as SPRICAGE and also authors, year of reference, number of sites and number of samples
SREFLIST	PREFLIST	Lists references for a user supplied author name in a format similar to a standard reference
SAUTHQRY	PAUTHQRY	Lists references for a user supplied author name and year. If the spelling of the author name is not known for certain, enter the spelling as far as is known. If the exact year is unknown, enter the decade by omitting the last digit of the year
ACTIVITY		Lists a summary of the paleomagnetic activities of a user supplied author
STIME	PTIME	Lists the 1989 geological time scale. TSCALE.DOC in the \PALEOMAG directory also contains the time scale and may be printed from MS-DOS or SQL*Plus via the HOST command.
SKEYS	PKEYS	Lists function keys for use with the customised form PALEOMAG
SJOURNAL	PJOURNAL	Lists the journal entry styles used in the REFERENCE table of the GPMDB
SINFO	PINFO	Lists explanations of the symbols used in the GPMDB
SALLVIEW	PALLVIEW	Lists, from the data dictionary, the code used to create all the database VIEWS you have created
SONEVIEW	PONEVIEW	Lists, from the database dictionary, the code used to create a your nominated VIEW
SINDEXES	PINDEXES	Lists, from the database dictionary, the index names, table and column indexes that you have created

Lines 4–6 set up the screen page definition:

SET NEWPAGE 1   – 1 line left blank at the top and bottom of the page
SET PAGESIZE 25   – 25 lines to the page
SET LINESIZE 80   – 80 characters to the line
SET SPACE 1   – 1 space is to be left between columns of output

Lines 7–12 determine what is displayed on the screen:

```
1    REM THIS IS A COMMAND FILE THAT RETRIEVES REFERENCES FOR A
2    REM USER SUPPLIED AUTHOR NAME AND OUTPUTS THEM TO THE SCREEN
3    REM IN A FORMAT SIMILAR TO A STANDARD PUBLISHED REFERENCE.
4    SET NEWPAGE 1
5    SET PAGESIZE 25;
6    SET LINESIZE 80;
7    SET SPACE 1;
8    SET VERIFY OFF;
9    SET ECHO OFF;
10   SET TERMOUT ON;
11   SET FEEDBACK ON;
12   SET PAUSE ON;
13   SET PAUSE "Press ENTER or RETURN to continue";
14   COLUMN AUTHORS FORMAT A79 WORD_WRAP;
15   COLUMN YEAR FORMAT 9999 JUSTIFY R
16   COLUMN JOURNAL FORMAT A32 WORD_WRAP;
17   COLUMN VOLUME FORMAT 9999 JUSTIFY R HEADING 'VOL';
18   COLUMN PAGES FORMAT A34;
19   COLUMN TITLE FORMAT A79 WORD_WRAP;
20   SELECT AUTHORS, YEAR, JOURNAL, VOLUME, PAGES, TITLE
21   FROM REFERENCE R, AUTHORS A
22   WHERE R.REFNO = A.REFNO
23   AND AUTHORS LIKE '%&AUTHORNAME%'
24*  ORDER BY YEAR,JOURNAL,VOLUME,PAGES;
```

Fig. 33. The SQL*Plus Command File SREFLIST.SQL.

SET VERIFY OFF   – the substitute for the variable in the SQL statement is not echoed

SET ECHO OFF   – the SQL commands in the command file are not echoed as they are executed

SET TERMOUT ON   – Output is directed to the screen

SET FEEDBACK ON   – The number of records returned by a query is displayed

SET PAUSE ON   – Output pauses at the end of each page

SET PAUSE "Press ENTER or RETURN to continue"
    – displays the message if PAUSE is ON

Lines 13–18 format the output to the screen

COLUMN column   – specifies the column to be formatted

FORMAT 9999   – displays an integer of up to 4 digits

FORMAT A79   – displays up to 79 characters

JUSTIFY Left, Center, Right
    – controls the placement of the heading over the column width allocated

HEADING 'string'   – allows the original column name to be changed

WORD_WRAP   – if the column contains more characters than allocated to it by FORMAT then the output wraps to the next line breaking on whole words.

Lines 19–23

SQL statement which selects the data from the database

Executing a .SQL Command File – the START command

The START command executes the contents of the specified command file. If no file extension is specified SQL*Plus assumes that it is .SQL. The file can contain any SQL*Plus or SQL command that can be run interactively. For example to execute the command file SREFLIST.SQL, enter:

SQL> START SREFLIST;

and SQL*Plus will prompt you for an author name.

NOTE: Once a SQL command (a SELECT statement) or a .SQL command file containing a SELECT statement is run, the current buffer is automatically reset to the SQL buffer regardless of its previous setting.

PREFLIST.SQL is a customised file that retrieves the same data from the database as SREFLIST.SQL. Output is spooled to an MS-DOS file called PREFLIST.LST in the work directory, C:\PALEOMAG and is automatically printed if your printer is switched on. The printer may not page correctly as the line where you entered the author name is included at the head of the file. To correct this, either edit PREFLIST.LST now in SQL*Plus using the line editor or later in MS-DOS using an editor or word processor in non-document mode.

Executing an MS-DOS Command – the HOST command

The HOST command allows an MS-DOS command to be executed without leaving SQL*Plus. Check with your DBA that this command is available on your system.

HOST system command – executes a single system command.
HOST with no command – gives the MS-DOS prompt C:\PALEOMAG >,
 then multiple system commands may be entered. To
 return to SQL type in EXIT.

Do not attempt to re-load ORACLE or re-log into SQLPLUS using the HOST command.

For example, to print the spooled file PREFLIST.LST from SQLPLUS enter:

SQL> HOST COPY PREFLIST.LST LPT1

NOTE: Our printer port is LPT1 and yours will most likely be the same. However it may be another LPT or COM port. In that case, use the name of your printer port in the above HOST command.

To print the spooled PREFLIST.LST from MS-DOS enter:

C:\PALEOMAG >PRINT PREFLIST.LST

or

C:\PALEOMAG >COPY PREFLIST.LST LPT1

If you re-run PREFLIST.SQL it will overwrite an existing PREFLIST.LST file. Remember to delete PREFLIST.LST when it is no longer needed.

To delete PREFLIST.LST from SQLPLUS enter:

SQL> HOST DEL PREFLIST.LST

To delete PREFLIST.LST from MS-DOS enter:

C:\PALEOMAG> DEL PREFLIST.LST

Custombuilt .SQL Files

Table XIV lists and briefly describes the customised files that are provided with the database. The files have a read only attribute so that they cannot be accidentally overwritten or deleted. Two versions of each customised file exists, one formatted for display on the screen (prefixed by an 'S') and the other formatted for output to the printer (prefixed by a 'P') with 66 lines to the page. This pages correctly for 11 inch paper and will need to be changed if you are using standard A4 paper. The files are spooled to the printer port LPT1. If your printer port is not LPT1 your DBA should have made the necessary changes to ensure that output spools correctly to your printer.

The code for the .SQL command files listed in Table XIV are given in Appendix 3.

Writing Query Results to an MS-DOS File – the SPOOL command

The SPOOL 'filename' command turns spooling on and writes displayed output to an MS-DOS file, filename .LST (or .LIS on some systems) in the current directory (the GPMDB work directory, C:\PALEOMAG). SPOOL OFF turns spooling off.

A listing of the code of any of the customised files may be obtained in the following manner:

On the screen:

SQL> SET BUFFER SCRATCH;
     GET filename;

On the printer:

SQL> SET BUFFER SCRATCH;
SQL> SPOOL LIST;
SQL> GET filename;
SQL> SPOOL OFF;
SQL> HOST COPY LIST.LST LPT1;
SQL> HOST DEL LIST.LST;

Output can be directly spooled to the printer without a .LST file being created.

SQL> SPOOL LPT1;
SQL> SET BUFFER SCRATCH;
SQL> GET filename;
SQL> SPOOL OFF;

Creating Your Own .SQL Command File – the SAVE command

The SAVE command saves the contents of the current buffer in an MS-DOS file in the current directory (the GPMDB work directory, C:\PALEOMAG). SQL*Plus gives the file an extension .SQL unless it is directed otherwise. Initially the current buffer is the SQL buffer. If the SQL command (a SELECT command) in this buffer is saved, a slash (/) is appended to the end of the file. Another buffer should be selected with the SET BUFFER command to SAVE a file that contains SQL*Plus and SQL commands.

We suggest that if you wish to make a variation of a customised file then make a copy of the file, edit it and save it under a new name. One way to do this is as follows:

SQL> SET BUFFER SCRATCH;
SQL> GET filename,

edit the file with the SQL Line Editor then:

SQL> SAVE newfilename;

NOTE: As the SQL*Plus command file is stored in a MS-DOS file called filename .SQL it can always be copied then edited using an editor or word processor in non-document mode.

Creating Your Own. SQL Command File – the INPUT command

The INPUT command allows text to be entered into the current buffer. It may be abbreviated to I.

INPUT no text – allows multiple lines to be added to the buffer after the current line (indicated by asterisk (*))
INPUT text – adds a single line to the buffer after the current line

If you wish to make your own .SQL command file then:

SQL> SET BUFFER SCRATCH;
SQL> I

Type in SQL*Plus commands;
Type in SQL SELECT statement;
To exit input mode press [Enter] twice

SQL> SAVE filename;

Then to run your command file:

SQL> START filename;

SQL command files can also be written using an editor or word processor in non-document mode. If you do this be careful to put a line feed at the end of the last line in the file before saving it. Failure to do this causes an "Input truncated to n characters" message to be issued.

Remember after you start the .SQL command file the current buffer automatically becomes the SQL buffer in order to run the SQL command. This means that if you discover an error in your command file code you must reset the current buffer to SCRATCH in order to view and edit the code with the SQL Line Editor.

SQL> SET BUFFER SCRATCH;
SQL> GET filename;

If you do not reset the current buffer L will list the contents of the SQL buffer (the current buffer) which contains the SQL SELECT statement but no SQL*Plus commands. You can edit this buffer but if you SAVE to the same file only the SELECT statement will be saved and the SQL*Plus commands will be lost.

If you find that one of the GPMDB files outputs in an unexpected format or unrelated top or bottom page titles or column names are given then previously started files may have set some SQL*Plus commands inappropriately. You may have inadvertently reset commands yourself. The simplest and usually the quickest way to rectify this problem is to exit SQL*Plus and log in again.

If you wish to protect your code from alteration or from being deleted then set the read only attribute in MS-DOS, or from SQL*Plus via the HOST command:

C:\PALEOMAG >ATTRIB +R filename .SQL

5. More Advanced SQL Queries

5.1. MULTIPLE TABLE JOINS

Several tables can be joined by linking them through their primary keys and the selected data combined into a single result table. The primary keys do not have to be the same for every table joined but can be used to link tables in a chain.

Suppose you want a list of the names of authors who have obtained rock magnetic data from their studies and the rock name involved for rocks aged between 0 and 2 M.a. (Quaternary). First format the output so that it will fit across the screen then enter the query (Figure 34):

SQL> COLUMN AUTHORS FORMAT A18 WORD_WRAP:
SQL> COLUMN ROCKNAME FORMAT A18 WORD_WRAP;
SQL> COLUMN ROCKMAG FORMAT A26 WORD_WRAP;

```
AUTHORS                ROCKNAME             ROCKMAG
------------------     ----------------     -------------------------

Chamalaun,F.H.         Reunion Island       Js-T(TM),SUSC
                       Group 1 Lavas

Doell,R.R., Cox,A.     Hawaiian Lavas       Js-T(CT 475-575C),AN
Mankinen,E.A.,         Volcanics, Long      Js-T(CT 525-570C),
Gromme,C.S.,           Valley Caldera       Jrs/Js(<0.2)
Dalrymple,G.B.,
Lanphere,M.A.,
Bailey,R.A.

Knight,M.D.,           Toba tuffs,          OP(TM low Ti),
Walker,G.P.L.,         Sumatra              Js-T(CT 400-550C),
Ellwood,B.B.,                               Jrs/Js,Hrc/Hc,AN
Diehl,J.F.

Geissman,J.W.          Quaternary           IRM,Js-T but no details
                       Volcanics, Valles
                       Caldera

Brown,L.,              Medicine Lake        OP(TM with exsolved ILM
Mertzman,S.A.          Lavas                lamellae,max Cl.3,no MAGH)

Bohnel,H.,             East Eifel           OP(TM oxidation state
Kohnen,H.,             Volcanics            varies)
Negendank,J.,
Schmincke,H-U.

       .

       .

Kono,M.                Usami Volcano        Js-T(CT 500-565C), X-Ray
                                            Thellier intensities


35 records selected.
```

Fig. 34.   Multiple table joins giving ROCKMAG data for rocks aged between 0 and 2 Ma.

SQL> SELECT AUTHORS, ROCKNAME, ROCKMAG
    FROM   AUTHORS A, ROCKUNIT RU, PMAGRESULT PMR
    WHERE A.REFNO = RU.REFNO
    AND   RU.ROCKUNITNO = PMR.ROCKUNITNO
    AND   ROCKMAG IS NOT NULL
    AND   HIGHAGE < 3;

Without the age condition this query will return every result in the GPMDB
that has rock magnetic data (there are over 1000). When entering SQL*Plus
commands such as the COLUMN command interactively they are in force until
the end of your current SQL*Plus session unless they are CLEARed. They are
not stored in the SQL buffer so a SQL prompt is issued after each command is
entered. Table XII lists the various commands that can be used to show the current
setting of a command.

(a)
```
COUNT(RESULTNO)
---------------
            73
```

(b)
```
                        MAX    MIN    MAX    MIN
     DATA SITES SAMPLES LAT    LAT    LONG   LONG
     ----- ----- ------- ----- ----- ------ ------
      73    879    3784  43.5  -35.0  151.4  -75.0
```

Fig. 35.   GROUP Functions based on author McElhinny. (a) Using the COUNT function. (b) Using the COUNT, SUM, MIN, MAX and revising the default format.

5.2. GROUP FUNCTIONS

A group function operates on several rows (a group – as specified in the where clause) or the entire table. The result will be one answer. Consequently when used in the SELECT clause, the only other items allowed in the SELECT clause are other group functions (that return a single result). The group functions are: COUNT, SUM, MIN, MAX, AVG and may operate on all the entries in a table, the distinct entries or on expressions.

For example, to determine how many results can be attributed to McElhinny whether or not he is the sole author of the paper, use the following SQL statement:

```
SQL> SELECT COUNT(RESULTNO)
     FROM    AUTHORS A, ROCKUNIT RU, PMAGRESULT PMR
     WHERE A.REFNO = RU.REFNO
     AND     RU.ROCKUNITNO = PMR. ROCKUNITNO
     AND     AUTHORS LIKE '%McElhinny%';
```

This query will take several minutes. Note that the title of the column RESULTNO shown in Figure 35(a) includes the GROUP function COUNT that operates on it.

Now extend this query also to find out how many sites and samples McElhinny has collected and the geographical range of his operations. At the same time the column headings can be changed as follows:

```
SQL> COLUMN COUNT(RESULTNO) FORMAT 9999 HEADING 'DATA';
SQL> COLUMN SUM(B) FORMAT 9999 HEADING 'SITES';
SQL> COLUMN SUM(N) FORMAT 9999 HEADING 'SAMPLES';
SQL> COLUMN MAX(SLAT) FORMAT 99.9 JUSTIFY C HEADING
         'MAX|LAT';
SQL> COLUMN MIN(SLAT) FORMAT 99.9 JUSTIFY C HEADING
         'MIN|LAT';
SQL> COLUMN MAX(SLONG) FORMAT 999.9 JUSTIFY C HEADING
         'MAX|LONG';
```

```
SQL> COLUMN MIN(SLONG) FORMAT 999.9 JUSTIFY C HEADING
     'MIN|LONG';
SQL> SELECT COUNT(RESULTNO), SUM(B), SUM(N), MAX(SLAT),
     MIN(SLAT), MAX(SLONG), MIN(SLONG)
     FROM   AUTHORS A, ROCKUNIT RU, PMAGRESULT
     PMR
     WHERE A.REFNO = RU.REFNO
     AND    RU.ROCKUNITNO = PMR.ROCKUNITNO
     AND    AUTHORS LIKE '%McElhinny%';
```

This query will now give a summary of McElhinny's paleomagnetic activities as recorded in the GPMDB (Figure 35(b)) You could leave out the COLUMN commands and accept the default formatting. However if you plan to write your own SELECT statements we suggest that you gain experience using the COLUMN commands as they can make the output from your queries much more readable. You may wish to try both ways to see the difference that formatting makes to the output. You could of course input the COLUMN commands and SELECT statement into a SQL command file using the method described in the previous section. This means any errors in entering the commands can be easily corrected by editing. Note that when entering SQL*Plus commands into a command file no SQL prompts are issued. The FORMAT option of the COLUMN command models the way numbers will appear in the query output by using a 9 to represent each digit. SQL*Plus adds one extra space to allow for the sign of the number. To print a HEADING on more than one line use a pipe (| – the symbol above the back slash \ on the keyboard).

```
SQL> SELECT COUNT(DISTINCT CONTINENT)
     FROM ROCKUNIT;
```

This query counts the 12 distinct continents and oceans.

```
SQL> SELECT COUNT(DISTINCT CONTINENT)
     FROM ROCKUNIT
     WHERE CONTINENT NOT LIKE '%Ocean';
```

This query counts only the 8 distinct continents.

5.3. SUMMARISING SEVERAL GROUPS OF ROWS

The GROUP BY Clause

When several rows have a common value in one or more columns they can be handled as a group. The GROUP BY clause returns one row for each group. The SELECT clause may only contain group functions and/or expressions specified in the GROUP BY clause. The WHERE or AND clause can be used in conjunction with the GROUP BY clause to eliminate rows before the grouping is performed. The general syntax is:

         SELECT   group-by-column(s) or functions
         FROM   table(s)

         [WHERE/AND row condition]

         GROUP BY   group-by-column(s);

To count the number of results in the GPMDB continent by continent enter the following query:

```
SQL> SELECT CONTINENT, COUNT(RESULTNO)
     FROM   ROCKUNIT RU, PMAGRESULT PMR
     WHERE RU.ROCKUNITNO = PMR.ROCKUNITNO
     GROUP BY CONTINENT;
```

This query will take several minutes (see Figure 36(a)). A condition could be added as given below.

The GROUP BY and the HAVING Clause

The HAVING clause is applied to the groups formed by the GROUP BY clause, and specifies the condition(s) the groups must satisfy. Only the groups which satisfy this condition are selected. The condition in the HAVING clause must be comprised of group functions or the expressions specified in the GROUP BY clause. Better performance is achieved by using the WHERE clause to eliminate rows before they are grouped. The general syntax is:

       SQL > SELECT   group by item, group functions
              FROM   table(s)
              [WHERE row condition]
              GROUP BY   column name[, column name]
              HAVING   condition (desired group characteristics)

Suppose a list of authors, together with a count of results, is required for those references each of which has more than 10 results.

```
SQL> COLUMN REFNO NOPRINT;
SQL> COLUMN AUTHORS FORMAT A48 WORD_WRAP;
SQL> COLUMN COUNT(RESULTNO) JUSTIFY C HEADING 'NO
     OF|RESULTS';
SQL> SELECT   A.REFNO, AUTHORS, COUNT(RESULTNO)
     FROM     AUTHORS A, ROCKUNIT RU, PMAGRESULT PMR
     WHERE    A.REFNO = RU.REFNO
     AND      RU.ROCKUNITNO = PMR.ROCKUNITNO
```

(a)

```
CONTINENT           COUNT(RESULTNO)
---------------     ---------------
Africa                      328
Antarctica                   71
Arctic Ocean                 16
Asia                        453
Atlantic Ocean               89
Australia                   241
Europe                      965
Greenland                    81
Indian Ocean                 22
North America              1232
Pacific Ocean                99
South America               188


12 records selected.
```

(b)

```
AUTHORS                                                   RESULTS
------------------------------------------------------   -------
Andriamirado,C.R.A.                                         16
Collinson,D.W., Runcorn,S.K.                                14
Creer,K.M.                                                  23
Du Bois,P.M.                                                12
Elston,D.P., Bressler,S.L.                                  12
Gose,W.A., Swartz,D.K.                                      11
Irving,E., Green,R.                                         11
Klootwijk,C.T.                                              13
          .
          .
          .
Nairn,A.E.M.                                               12
Nairn,A.E.M.                                               12
Palmer,H.C.                                                12
Perroud,H., Bonhommet,N.                                   11
Piper,J.D.A.                                               27
Piper,J.D.A.                                               17
Piper,J.D.A.                                               11
Stearn,J.E.F., Piper,J.D.A.                                17
VandenBerg,J., Klootwijk,C.T., Wonders,A.A.H.              13
Zhang,H., Zhang,W.                                         34


22 records selected.
```

Fig. 36. The GROUP BY Clause giving the number of results for each continent. (a) The simple GROUP BY. (b) Conditional GROUP BY using the HAVING clause to give authors with more than 10 results in any given paper.

GROUP BY A.REFNO, AUTHORS
HAVING   COUNT(RESULTNO) > 10
ORDER BY AUTHORS,

This query will take several minutes (see Figure 36(b)). There are several things to note about this query. As it is not necessary to list REFNO, the NOPRINT option of the COLUMN command has been used. However, REFNO needs to be selected as it is required to ensure that the GROUP BY clause operates correctly. If the grouping was only by AUTHORS and the identical author name appeared more than once, then the GROUP BY would treat that author(s) as a group and the number of results would be added together. This is not the intent of the query. The REFNO is unique and only this, as a group, can force each individual reference to be treated separately. As can be seen in Figure 36(b) 'Piper' is listed three times. Without grouping by REFNO he would have appeared only once. This is a very complex query and requires a sophisticated understanding of SQL syntax. As more complex queries are attempted, it may be necessary to devise tests to be sure that the query is actually operating as intended.

5.4. SUBQUERIES – THE IN CLAUSE

The WHERE clause of one query may contain another query, called a subquery. The subquery is used dynamically to build a search condition for the main query. ORACLE processes subqueries before main queries because it needs the result of the subquery to find the result of the main query. General syntax:

```
SQL> SELECT column1, column2, column3, . . .
     FROM   table
     WHERE column = (SELECT column
                     FROM   table
                     WHERE condition);
SQL> SELECT AUTHORS
     FROM   AUTHORS
     WHERE REFNO IN (SELECT REFNO
                     FROM   ROCKUNIT
                     WHERE PLACE LIKE '%France%');
```

This query will return the authors of papers who have made studies in France or French territories around the world. Information comes from one table but the selection criteria operates on another table. The reference numbers for the French data are determined in the subquery and passed to the main select statement that returns the authors.

There are references in the GPMDB that are cross-referenced by other studies that have results in the database, but themselves contain no results. Suppose you wish to obtain a listing of these. There is more than one way to do this and following example is a query NOT to try! The use of a subquery can be a disaster if you don't think carefully about what your query will actually do.

```
SQL> SELECT REFNO
     FROM   REFERENCE
     WHERE REFNO NOT IN (SELECT REFNO
                         FROM   ROCKUNIT);
```

This looks simple enough but think about what happens. The bracketed SELECT (the sub-query) is carried out first. ALL reference numbers will be selected from

ROCKUNIT, about 4000 of them! Then each of the 2000 or so REFERENCEs must be compared with them. This is 4000×2000 or 8 million comparisons that have to be made! Suppose these comparisons can be done at about 500 a second, then 8 million operations will take about 4.5 hours. A much quicker and more efficient way to get the same information is by using an OUTER JOIN (Section V.5.5 below).

Some Notes on using Subqueries:

1. The comparison operator in the where clause is not restricted to =.
2. The subquery is enclosed in ( )'s.
3. The subquery query may retrieve values from a different table to the outer query.
4. Subqueries have many features and some restrictions. Consult the 'SQL*Plus User's Guide' for more information and examples.

5.6. THE OUTER JOIN

In an *EQUI-JOIN*, for each row selected in one table, there is at least one corresponding row in each of the tables joined. If a row in one of the tables does not satisfy the join condition, then that row will not appear in the query result. To ensure that the unmatched rows are retrieved by a query use an *OUTER JOIN*. The *OUTER JOIN* specifies that where, for one table, there is no corresponding entry in the other table a null value will be returned. The (+) indicates the table where the extra nulls must appear.

```
SQL>  SELECT R.REFNO, RU.REFNO
      FROM   REFERENCE R, ROCKUNIT RU
      WHERE R.REFNO = RU.REFNO (+)
      AND    RU.REFNO IS NULL;
```

This is the information we sought in the last query, however, this command will take about a minute instead of 4.5 hours.

NOTE: A table can be outer joined to, at most, one other table.

5.6. JOIN VERSUS SUBQUERY

In general, in order to decide whether your query requires a join or a subquery, look at where the data will be retrieved from. Use a join when data in the main SELECT clause comes from two or more tables. Use a subquery when all the data comes from ONE table and the nested select simply assists in choosing the rows that are to appear.

5.7. Optimising SQL queries

Some ORACLE system parameters can be fine-tuned but this only improves performance by up to about 5%. Large improvements in performance of up to 300% can be achieved by optimising SQL queries.

Writing Efficient SQL Code

The following list offers some guidelines for writing more efficient SQL code.

1. Avoid the use of * in queries as access is much faster when the columns are named. For example, do not use SELECT * or COUNT (*).
2. Specify tablename.columnname in table joins. ORACLE then does not need to look up the data dictionary for the table name.
3. Use brief table name aliases when tables are joined. For example:

```
SQL> SELECT ROCKNAME, ROCKTYPE, SLAT, SLONG, B, N, DEC,
            INC, PLAT, PLONG
     FROM   PMAGRESULT PMR, ROCKUNIT RU
     WHERE PMR.ROCKUNITNO = RU.ROCKUNITNO
```

4. Specify the names of the columns that data are to be INSERTed INTO.
5. Rephrasing a query can improve performance. For example, this query removes the groups not required after the groups have been made:

```
SQL> SELECT    CONTINENT, AVG(LOWAGE), AVG(HIGHAGE)
     FROM      ROCKUNIT
     GROUP BY CONTINENT
     HAVING    CONTINENT != 'North America';
```

It is more efficient to remove the rows not required before making the groups. So rephrase the query to eliminate the rows not required before the groups are made.

```
SQL > SELECT    CONTINENT, AVG(LOWAGE), AVG(HIGHAGE)
      FROM      ROCKUNIT
      WHERE     CONTINENT != 'North America'
      GROUP BY CONTINENT;
```

When using a GROUP BY early elimination of rows can optimise the query.
6. If there are a number of selection conditions to be applied in a query it may be faster to list these first to restrict the rows returned before joining the tables. For example:

```
SQL > SELECT AUTHORS, YEAR, JOURNAL, VOLUME, PAGES
      FROM   AUTHORS A, REFERENCE R
      WHERE AUTHORS LIKE 'McElhinny%'
```

AND   JOURNAL = 'J.Geophys. Res.'
AND   A. REFNO = R.REFNO;

restricts the rows returned by the query firstly to those with first author McElhinny, then further restricts the rows to those in J.Geophys.Res, and lastly joins the A and R tables together. If the join condition were given first then all rows in the tables would have been joined before any rows were eliminated.

Writing SQL Code to Use Indexes

Indexes are compact tables that can be created to increase the speed of data retrieval. They can be used to enforce uniqueness (for example unique indexes on primary keys). They may be defined on a single column or jointly on several columns (concatenated indexes) of a table and may be compressed or non-compressed. Once created, ORACLE automatically maintains and uses them. See Section III.7 (ORACLE Indexes) for a more detailed discussion of indexes. Table VII lists the indexes supplied with the GPMDB. These are mainly unique compressed indexes on the primary keys of the GPMDB tables.

The way the SQL query is written will determine whether the ORACLE optimiser can use available indexes. The ORACLE optimiser analyses a SQL statement and decides how to run it as efficiently as possible based on some default assumptions about the data and on the indexes that are available to achieve this. For an index to be used in a SQL statement:

- it must contain a WHERE clause and
- the indexed column is the one used to include or exclude rows returned by a query.

Joins should always be indexed, as indexes can dramatically improve the performance of table joins. Non-indexed joins involve whole table scans and these are very slow, especially for multiple joins of large tables. In a join the tables are merged by matching information from one table , the 'driving table', to information in the other tables. If the join-conditions of the tables are indexed and the indexes are of equal status, the ORACLE optimiser chooses the last table in the FROM clause as the 'driving table'. To optimise the efficiency of the join:

- make the table with smallest number of qualified rows LAST in the FROM clause
- when joining more than two tables, list the join condition that returns the smallest number of rows LAST in the WHERE clause.

The GPMDB has been designed so that each table has either a single or a joint primary key that uniquely identifies every row in the table. Tables in the GPMDB are always joined on these keys, so unique compressed indexes have been created on them (see Table VII). We have run timing tests on table joins using different 'driving tables' and listing the join conditions in different order. We found no significant difference in the speed of the join except where no indexes were used. However the above discussion suggests that for tables of very different sizes this may not be the case.

Existing indexes will not be used by the ORACLE optimiser in the following cases:

- With the DISTINCT operator – ORACLE creates a temporary table and a temporary index on that table then finds the distinct entries in the specified column.
- With LIKE '%string' – here a whole table scan is used.
- With the GROUP BY clause – ORACLE uses a temporary table to store the groups as they are made.
- With a column that is modified by an ORACLE function or arithemetic operator – where possible, re-arrange the query so that an index can be used.
- If the query contains an *OUTER JOIN*.
- If the optimiser decides that using the index will not improve performance.
- NULL values are not indexed. For example, an index on ROCKMAG cannot be used in the following query:

      SQL> SELECT some data
            FROM   PMAGRESULT
            WHERE ROCKMAG IS NULL;

The next query will not use an index on ROCKMAG as ORACLE makes the assumption that the majority of records in the column contain data so a whole table scan is executed.

      SQL> SELECT some data
            FROM   PMAGRESULT
            WHERE ROCKMAG IS NOT NULL;

The following query will use an index on ROCKMAG to find the NOT NULL values and is appropriate when a significant number of records do not have ROCKMAG data.

      SQL> SELECT some data
            FROM   PMAGRESULT
            WHERE ROCKMAG > '';

- NOT EQUAL clauses do not use indexes. ORACLE makes the assumption that the number of rows selected is significantly greater than the number of rows rejected and so a full table scan is executed. Try to avoid the use of NOT. The following negations are converted by the optimiser.

NOT >   is converted to < =
NOT > =   is converted to <
NOT <   is converted to > =
NOT < =   is converted to >

When multiple indexes are available the ORACLE optimiser must choose which indexes to use:

- If the indexed columns are used with range conditions in the query, only the first index will be used. For example, if there were non-unique indexes on each of the B and N columns of PMAGRESULT then the following SQL query:

      SQL> SELECT   some data
           FROM   PMAGRESULT
           WHERE   B > 10
           AND   N > 20;

will use only the index on B to identify qualified rows. These would then be checked directly to identify the rows that satisfy the condition on N.
- Unique indexes are more efficient than non-unique indexes as there are no duplicate entries. If both types are available ORACLE will choose the unique index to select data and ignore the non-unique index. For example, if there were a non-unique index on LOWAGE and a unique index on the primary key ROCKUNITNO in the ROCKUNIT table:

      SQL> SELECT   some data
           FROM   ROCKUNIT
           WHERE   LOWAGE = 100
           AND   ROCKUNITNO = 1000;

will use the ROCKUNIT index and, if a record is found, will examine it directly without using the LOWAGE index to see if the LOWAGE is 100.

- Multiple indexes may not be used for performance reasons. A query with two or more conditions in the WHERE clause can use multiple indexes if the indexes are on:
  - columns compared to a constant (equalities); then up to 5 indexes will be merged
  - columns that are non-uniquely indexed; then up to 5 indexes may be merged
  - columns from the same table
  - mixture of unique and non-unique, only the unique indexes will be used
  - columns used in a mixture of equality and range comparisons; only the indexes on columns of equalities will be used
  - columns used in range comparisons; only the first index will be used.

6. Using SQL*Plus Commands to Operate on Your Own Data

Users of the GPMDB do not own any of the GPMDB tables or views. As owners of the tables and views we have restricted your privileges on them to SELECT only. You are not permitted to change existing data entries (UPDATE), add data entries (INSERT) or remove data entries (DELETE). This is to protect the integrity of the GPMDB. You may wish to develop special data sets for your own purposes on which you may want to have all of the above privileges. This can be done by creating your own tables from the GPMDB. As the owners of these tables, you will then have all of the available privileges on them. Changes you make to entries in your tables do not change the GPMDB in any way.

If you work through this section and CREATE TABLES, VIEWS, SYNONYMS and GRANT privileges remember to DROP the TABLES, VIEWS, SYNONYMS and to REVOKE the privileges when you are finished with them.

If you aim to be an advanced SQL*Plus user then we highly recommend that you work through the SQL*Plus Tutorial in the 'SQL*Plus User's Guide'. To do this you will need to load the sample tables into the GPMDB work directory, C:\PALEOMAG.

> C:\PALEOMAG >DEMOBLD youruserID/password

DEMOBLD creates a LOGIN.SQL file that contains appropriate commands to ensure the tutorial operates correctly. Your original LOGIN.SPL is re-named LOGIN.OLD.

When you have finished the tutorial remove the sample tables as follows:

> C:\PALEOMAG >DEMODROP youruserID/password

DEMODROP will delete the LOGIN.SQL created by DEMOBLD and re-name LOGIN.OLD to LOGIN.SQL.

6.1. CREATING, DROPPING AND ALTERING TABLES

The CREATE TABLE Command

The CREATE TABLE command creates a table, defines its columns and attributes. There are two forms of the command.

1. The CREATE TABLE command that redefines the column definitions

> SQL> CREATE TABLE tablename (column1name column1type column1size
> column1specification, column2name . . . .);

The first form of the command creates the table definition in ORACLE's data dictionary. Data entries may then be made using the one of the forms of the INSERT command described in Section V.6.2 (Changing Data in your Tables).

tablename
- is the name of the table
- must be unique to a user
- must not be an ORACLE reserved word.
- (See the 'SQL Language Reference Manual' for a list of reserved words)
- must begin with a character
- may be up to 30 alphanumerics in length
- may be upper, lower or mixed case
- must match the case entered in the database exactly
- may not contain spaces, instead use the underscore _

columnname
- is a name of a column in a table
- does not have to be unique to the user but must be unique to the table

column type and size – columns in the GPMDB may be:

- alphanumeric and up to 240 characters long (e.g. CHAR(60) stores up to 60 alphanumerics)
- numeric in type and size, for example:
  NUMBER(4)   a numeric of size up to 4 digits
  NUMBER(5, 2) a real number of up to 5 significant digits 2 of which are after the decimal point

column specification – columns in the GPMDB may be:

- NULL, entries into the column are optional
- NOT NULL, entries into the column are mandatory

If you wish to create a compacted reference table that would contain your references and lists AUTHORS, YEAR, JOURNAL, VOLUME and PAGES enter:

SQL> CREATE TABLE MYREF (AUTHORS CHAR(50) NOT NULL,
                         YEAR NUMBER(4) NOT NULL,
                         JOURNAL CHAR(30) NOT NULL,
                         VOLUME NUMBER(4),
                         PAGES CHAR(11) NOT NULL);

"Table created."

The message "Table created" will be displayed when the CREATE TABLE command successfully completes. Notice that the column size attributes are different from those originally specified in the AUTHORS and REFERENCE tables (see the description of the tables in Appendix 1). This form of the CREATE TABLE command is the only way to redefine the original column specifications. The table definition for MYREF has been entered into the database but it contains no rows. These must be entered using an INSERT command described below in Section 6.2.

2. The CREATE TABLE   command that includes a query

SQL> CREATE TABLE   tablename [(alias1, alias2, . . . , aliasN)]
       AS SELECT   columns (same number of columns as aliases listed above)
       FROM   *   tablename(s)
       WHERE   optional conditions
       AND/OR   optional conditions,

This second form of the command creates the table definition in the data dictionary then inserts the rows selected by the query into the table. This imposes some restrictions on the columns of the new table. The new table may rename the columns but not redefine them. That is, the definition of the column attributes used in the original tables, cannot be changed. Changing the names of the columns is optional. If you do not wish to change the names of the columns in the new table they need not be listed in the CREATE TABLE command. The ORDER BY clause may not be used in the SELECT clause of the CREATE TABLE command. If you wish to order the rows in the table then you musrt do so when you query it (i.e. SELECT FROM it).

If you wish to create table MYREF and at the same time enter the rows into the table:

SQL> CREATE TABLE MYREF
       AS SELECT AUTHORS, YEAR, JOURNAL, VOLUME, PAGES
       FROM   AUTHORS A, REFERENCE R
       WHERE A.REFNO = R.REFNO
       AND   AUTHORS LIKE '%yourname%';

"Table created."

This enters the table definition into the database and enters the rows which match the condition in the query. The column names for the new table do not appear in the CREATE TABLE command so the column names assigned will be those listed in the SELECT statement (i.e. those in the original tables). The definition of the column attributes cannot be changed in this form of the CREATE TABLE command.

You may wish to create a table containing some basic paleomagnetic information for the redbeds in the database.

SQL> CREATE TABLE REDBED (ROCKNAME, PLACE, CONTINENT,
                         SITES,   SAMPLES,   SITELAT,
                         SITELONG, DEC, INC, POLELAT,
                         POLELONG)

    AS SELECT ROCKNAME, PLACE, CONTINENT, B, N, SLAT,
        SLONG, DEC, INC, PLAT, PLONG
    FROM   ROCKUNIT RU, PMAGRESULT PMR
    WHERE RU.ROCKUNITNO = PMR.ROCKUNITNO
    AND   ROCKTYPE LIKE '%redbeds%';

"Table created."

Notice that the names of some of the columns in REDBED have been changed from the original names listed in the SELECT statement. However the column attributes have not been changed. The number of columns listed in the CREATE TABLE command and the SELECT statement must be the same for the table to be successfully created.

Suppose you want to create a table containing information about all studies with a positive fold test:

SQL> CREATE TABLE FOLDTEST
    AS SELECT AUTHORS, YEAR, ROCKNAME, PLACE,
        LOMAGAGE, HIMAGAGE, PLAT, PLONG,
        PARAMETERS
    FROM   AUTHORS A, REFERENCE R, ROCKUNIT RU,
        PMAGRESULT PMR, FIELDTESTS FT
    WHERE A.REFNO = R.REFNO
    AND   A.REFNO = RU.REFNO
    AND   RU.ROCKUNITNO = PMR.ROCKUNITNO
    AND   PMR.RESULTNO = FT.RESULTNO
    AND   TESTS LIKE '%F+%'

"Table created."

The DROP TABLE Command

The DROP TABLE command drops one of your tables from the database. You cannot drop a table created by another user (i.e. one you do not own). All the rows in the table are deleted, freeing the disk space they occupied. Then the table definition is removed from the database. Tables that are no longer required should be dropped from the database to avoid filling the limited database file space that is available. Once you have issued the DROP TABLE command you cannot recover the table unless it has been backed up. If you wish to retain the information in the table you can either:

1. Store it in an MS-DOS file on the hard disk or a floppy diskette.

    SQL> SPOOL filename.LST;
    SQL> SET HEADING OFF;
    SQL> SELECT * FROM tablename;

SQL> SPOOL OFF;

This will spool the contents of the table to an MS-DOS file with the extension .LST (you can choose another extension) in the current directory (usually the paleomagnetic work directory, C:\PALEOMAG).

2. Back up the data and table definition using the EXP utility. The file can be stored on the hard disk or a floppy diskette. You will not be able to view the data as it is stored in an ORACLE compressed form. To view the data use the IMP utility to restore the table to the database. See your DBA for instructions on how to do this.

The DROP TABLE command has the syntax:

SQL> DROP TABLE tablename;

When a table is dropped indexes and privileges on it are also dropped. Synonyms, views and queries on the table become invalid and should be dropped or revised.

To DROP the table FOLDTEST enter:

SQL > DROP TABLE FOLDTEST;

The ALTER TABLE Command

A table can be altered in two ways.

1. Changing a Column Definition (ALTER TABLE MODIFY)

The ALTER TABLE MODIFY command has the following syntax:

SQL> ALTER TABLE tablename
         MODIFY (columnname   columntype   columnsize   columnspecification
         [,other columns to be modified]);

A column • can always be increased in size, the number of decimal places changed or the specification changed from NOT NULL to NULL

        • can only be decreased in size or the data type changed if it has been specified NOT NULL

        • specification can only be changed from NULL to NOT NULL if there are no null values in it

More than one column can be altered.

Suppose the JOURNAL field in MYREF is found to be too small. Then widen the column as follows:

SQL> ALTER TABLE MYREF
         MODIFY (JOURNAL CHAR(100));

The JOURNAL column may now contain up to 100 alphanumerics.

2. Adding a New Column to the Table (ALTER TABLE ADD)

The ALTER TABLE ADD command has the following syntax:

SQL> ALTER TABLE tablename
    ADD (columname columntype column size [, other columns to be added)
- New columns are added to the right of the last existing column in the table
- All fields in the new column are NULL and may not be defined as NOT NULL unless the table contains no rows

More than one column can be added to a table.

You may decide that the MYREF table should contain a TITLE and REFNO column so it can be linked to other tables in the GPMDB.

SQL> ALTER MYREF
    ADD (TITLE CHAR(240), REFNO NUMBER(4));

The TITLE and REFNO columns are added to the table to the right of the PAGES column and are now both NULL unlike the original definition in the REFERENCE table where they were specified NOT NULL. Entries into the table can only be made one at a time using the UPDATE command described below in Section V.6.2. below. If you want to add data to one of your tables and the data already exist in the database it is better to DROP the table and re-CREATE it with the extra columns. The ALTER TABLE ADD command is useful if you want to add new data that is not already in the GPMDB to one of your tables.

6.2. CHANGING DATA IN YOUR TABLES

There are three commands that can be used to modify the contents of your tables.

The UPDATE Command and the SET clause

The UPDATE command changes one or more values in a row or rows of your table. The syntax is as follows:

SQL> UPDATE   tablename
        SET   column1name = value1, column2name = value2, . . .
    WHERE   condition;
The SET   clause specifies the columns in the table that are to be updated and the new values that are to be entered. If a value is alphanumeric it must be enclosed in single quotes.
The WHERE   clause gives a condition that specifies the rows of the table that are to be updated. If no condition is given all the rows in the table will be updated.

Suppose you want to update your REDBED table and change the locality for all the studies of the Chugwater Formation in Wyoming to the same value 43N, 108W.

SQL> UPDATE REDBED
SET   SITELAT = 43.0, SITELONG = − 108.0
WHERE   ROCKNAME LIKE '%Chugwater%'
AND   PLACE LIKE '%Wyoming%';

"n records updated."

The INSERT INTO Command

The INSERT INTO command has two forms.

1. It may be used to add one row to your table by specifying the values. The general syntax is:

SQL> INSERT INTO tablename [(column1, column2, . . .)]
VALUES (value1, value2, . . .);

In VALUES you may list:

- values for all columns in the table, and must do so if they have been specified NOT NULL or are not listed in the INSERT INTO
- only values for columns specified NOT NULL.

The values to be inserted must be listed in the same order as the columns occur in the table.

2. It may be used to add a number of rows to a table by issuing a query. The general syntax is:

SQL> INSERT INTO tablename [(column1, column2, . . .)]
SELECT column1, column2, . . .
FROM   tablename(s)
WHERE   condition
AND/OR optional conditions;

- All the rows selected by the query are inserted into the table.
- The columns specified by the SELECT statement are matched to those listed in the INSERT INTO.
- If no columns are listed in the INSERT INTO command the columns listed in the SELECT statement must match all the columns in your table.

If you created MYREF using the first form of the CREATE TABLE command, which allows the columns to be redefined, then data can be inserted into the tables as follows:

SQL> INSERT INTO MYREF
SELECT AUTHORS, YEAR, JOURNAL, VOLUME, PAGES
FROM   AUTHORS A, REFERENCE R

WHERE A.REFNO = R.REFNO
AND   AUTHORS LIKE '%yourname%';

"n records created."

No columns are listed for MYREF so each column in the SELECT statement must match one in your table and they must be listed in the same order to ensure the data are inserted into your table correctly.

The DELETE Command

The DELETE command can delete one or more rows from one of your tables. If no WHERE condition is specified ALL the rows in your tables will be deleted and cannot be recovered unless your table has been backed up. The table definition is still recorded in the database. The general syntax is:

SQL> DELETE FROM tablename
            WHERE condition;

The rows meeting the condition in the WHERE clause are deleted. The condition can include a query.

For example to DELETE all pre-1975 references from your table MYREF enter:

SQL> DELETE FROM MYREF
            WHERE YEAR < 1975;

"n records deleted."

6.3. CREATING AND DROPPING VIEWS

The CREATE VIEW Command

The CREATE VIEW command specifies a SQL query that defines a VIEW of the GPMDB. A VIEW contains no rows of its own. Therefore you can VIEW the GPMDB but cannot change the entries. The general syntax is:

SQL> CREATE VIEW viewname [(alias1, alias2, . . .)]
            AS SELECT   column(s)
                 FROM   table(s) and/or view(s)
                 WHERE   optional conditions
                 AND/OR further optional conditions;

Viewname   • has the same restrictions as for table names
Alias   • if specified, become the names of the columns in the view
                  • if not specified, the column names will be the same as those of the
                    table or view in the FROM clause of the SELECT statement

The ORDER BY clause cannot be used in the SELECT clause of the CREATE

VIEW command. If you want the rows in a VIEW ordered then you must do so in a query (SELECT FROM) on the VIEW.

Create a VIEW that allows you to list all the journals and books referenced in the GPMDB:

    SQL> CREATE VIEW GPMDB_JOURNAL
            AS SELECT DISTINCT JOURNAL
            FROM   REFERENCE;

"View created."

A VIEW can be queried in the same way as a table. To list rows in GPMDB_ JOURNAL in alphabetical order enter:

    SQL> SELECT *
            FROM GPMDB_JOURNAL
            ORDER BY JOURNAL;

Suppose you want a list of the references in the GPMDB for the Mesozoic for part of the west coast of North America. This can be done by creating a VIEW specifying a latitude/longitude window and an age window:

SQL> CREATE VIEW MESOLIT
        AS SELECT AUTHORS, YEAR, JOURNAL, VOLUME, PAGES, TITLE
            FROM   ROCKUNIT RU, AUTHORS A, REFERENCE R
            WHERE RU.REFNO = A.REFNO
            AND   A.REFNO = R.REFNO
            AND   HIGHAGE > 65
            AND   LOWAGE < 245
            AND   RLAT BETWEEN 40.0 AND 55.0
            AND   RLONG BETWEEN −135.0 AND −120.0;

"View created."

Notice the way that the age window has been specified. This is to ensure that references for rocks with ages that overlap into the window from either end are selected. Also notice that for the BETWEEN operator that the smaller number is given first.

Suppose you want to VIEW some basic information about the limestones in the GPMDB.

SQL> CREATE VIEW LIMESTONE
        AS SELECT ROCKNAME, PLACE, CONTINENT, B, N, SLAT, SLONG, DEC, INC, PLAT, PLONG
            FROM   ROCKUNIT RU, PMAGRESULT PMR

WHERE RU.ROCKUNITNO = PMR.ROCKUNITNO
AND   ROCKTYPE LIKE '%limestone%';

"View created."
To query the VIEW:

SQL> SELECT *
FROM LIMESTONE
WHERE PLACE LIKE '%Canada%';

A VIEW of a VIEW can be created:

SQL> CREATE VIEW AFRICALST
AS SELECT ROCKNAME, PLACE, B,N, SLAT, SLONG
FROM   LIMESTONE
WHERE CONTINENT = 'Africa';

"View created."

The DROP VIEW Command

The DROP VIEW command deletes the VIEW definition from the database. The syntax is:

SQL> DROP VIEW viewname;

Synonyms, views and queries based on the view become invalid and should be revised or dropped.
   For example to DROP VIEW GPMDB_JOURNAL enter:

SQL> DROP VIEW GPMDB_JOURNAL;

"View dropped."

6.4. TABLES VERSUS VIEWS

CREATE a TABLE only when you wish to modify data that you have selected from the GPMDB. When a table is created the table definition and all the rows are stored in the database. This may occupy large amounts of database file storage space. Currently there are 8 Megabytes of database file space reserved on your hard disk. More than half of this space is taken up by the GPMDB. If user created tables proliferate, this space may be filled. In this case ORACLE will issue an "out of space in partition" message. If this occurs see your DBA.
   If you only wish to look at data in the GPMDB it is more efficient to CREATE a VIEW. When a view is created only the view definition is stored in the database. Data will be accessed more quickly through a view than a table as the SQL code which creates the view is parsed before it is stored in the database. In some

circumstances ORACLE will permit modification of the database tables underlying a view. You cannot modify data in the GPMDB through a view.

6.5. RENAMING TABLES AND VIEWS

The RENAME command re-names a table or view that you have created. The syntax is:

> SQL> RENAME   table or view name
>   TO   new table or view name;

For example to re-name the view AFRICALST enter:

> SQL> RENAME AFRICALST
>   TO   AFLIST;

"Rename succeeded."

6.6. CREATING AND DROPPING SYNONYMS

The CREATE SYNONYN Command

The CREATE SYNONYM command creates a synonym for a table or view name. The synonym can then be used in place of the table or view name. If you wish to refer to another user's table or view you must prefix it with the name of the user who created it followed by a full stop (.). You must also have the SELECT privilege on it (see Section V.6.7 – Granting and Revoking Priveleges). The syntax is:

> SQL> CREATE synonymname
>   FOR userID.tablename or userID.viewname;

For example if MIKE had a table in the database called LOTSADATA (it is not actually in the database) on which you have the SELECT privilege then to SELECT data you could enter:

> SQL> SELECT *
>   FROM MIKE.LOTSADATA;

To CREATE SYNONYM for MIKE.LOTSADATA you could enter:

> SQL> CREATE SYNONYM MDATA
>   FOR MIKE.LOTSADATA;

"Synonym created."

Now you can use the SYNONYM in place of the userID.tablename or userID.viewname in almost all clauses of SQL commands and in SQL*Plus commands. For example you could then:

SQL> SELECT *
FROM MDATA;

We have created a special synonym called a public synonym for all the GPMDB TABLES and VIEWS so that they can be referred to without the userID prefix. However only the DBA who owns the TABLES and VIEWS can CREATE PUBLIC SYNONYMs on them.

The DROP SYNONYM Command

The DROP SYNONYM command may be used to drop one of your synonyms from the database. You cannot DROP SYNONYMs created by another user. The general syntax is:

SQL> DROP SYNONYM synonymname;

For example had you been able to CREATE SYNONYM MDATA to remove it from the database enter:

SQL> DROP SYNONYM MDATA;

"Synonym dropped."

6.7. GRANTING AND REVOKING PRIVILEGES

The GRANT Command

The GRANT command allows you to grant access privileges on your tables and views (those created by you) to other users. The syntax is:

SQL> GRANT   privilege(s)
      ON   tablename or viewname
      TO   userID [,other userIDs];

The privileges that you can grant are:

SELECT	data from a table or view
UPDATE	values in a table
INSERT	rows into a table
DELETE	rows from a table
ALTER	column definitions in a table
INDEX	columns of a table
ALL	grants all privileges held by you
WITH GRANT OPTION	allows the user to pass on their privileges to other users.

Suppose MIKE and JO are authorised users (they are not) of the database and you wish to GRANT them some privileges.

SQL> GRANT SELECT, UPDATE
    ON   MYREF
    TO   MIKE, JO,

"Grant succeeded."

They could now select data from MYREF and update it but could not insert or delete data.

You may want to allow MIKE to SELECT data using your view MESOLIT and to have the ability to pass this privilege on:

       SQL> GRANT SELECT
            ON   MESOLIT
            TO   MIKE
            WITH   GRANT OPTION;

       "Grant succeeded."

We recommend the you grant only the SELECT privilege on views.

If you attempt to perform an unauthorised operation on a table or view, SQL* Plus will not allow the operation to take place.

- if you have only the SELECT privilege on a table or view and you attempt another operation ORACLE will issue an "insufficient privileges" message.
- if you have no privileges on a table or view that exists and you attempt to access it ORACLE will issue a "table or view does not exist" message.

The REVOKE Command

The REVOKE command removes one, several or all of the privileges that you have granted to a user on your table or view. The syntax is:

       SQL> REVOKE   privilege(s)
            ON   table or view name
            FROM   userID [, other users],

For example to REVOKE the UPDATE privilege on MYREF from MIKE and JO enter:

       SQL> REVOKE UPDATE
            ON   MYREF
            FROM   MIKE, JO;

"Revoke succeeded."

To REVOKE the SELECT privilege on view MESOLIT from MIKE:

       SQL> REVOKE SELECT
            ON   MESOLIT
            FROM   MIKE,

"Revoke succeeded."
This will also REVOKE the SELECT privilege from any user MIKE may have granted it to using the WITH GRANT OPTION.

References

Codd, E. F.: 1970, 'A Relational Model of Data for Large Shared Data Bases', *Communications of the ACM* **13**, 77–387.

Irving, E.: 1964, *Paleomagnetism and its Application to Geological and Geophysical Problems*, Wiley, New York.

McElhinny, M. W. and Lock, J.: 1990a, 'IAGA Global Palaeomagnetic Database', *Geophysical Journal International* **101**, 763–766.

McElhinny, M. W. and Lock, J.: 1990b, 'Global Palaeomagnetic Database Project', *Physics of the Earth and Planetary Interiors* **63**, 1–6.

McFadden, F. R. and Hoffer, J. A.: 1985, *Data Base Management*, Benjamin/Cummings, Menlo Park, California.

McFadden, P. L. and McElhinny, M. W.: 1990, 'Classification of the Reversals Test in Palaeomagnetism', *Geophysical Journal International* **103**, 163–169.

Perry, J. T. and Lateer, J. G.: 1989, *Understanding ORACLE*, Sybex, San Francisco, California.

Van der Voo, R. and McElhinny, M. W.: 1988, 'Global Paleopoles off to a Good Start', *EOS, Transactions of the American Geophysical Union* **69**, 819.

Acknowledgements

An international project like this one can only be undertaken with the enthusiastic participation of colleagues and agencies in many countries. These agencies are listed in the Introduction (Section I.1) and we gratefully acknowledge the financial support received either directly or indirectly. In particular we should especially like to thank the following colleagues who assisted with raising the necessary funds. They are Jim Briden, Peter Cook, Vincent Courtillot, Aleksei Khramov, Masaru Kono, William Lowrie, Phil McFadden, Heinrich Soffel, Rob Van der Voo and Hans Zijderveld.

Several colleagues acted as trial users of early versions of the database and we thank Chris Chouker and Rob Van der Voo in particular for their painstaking editing of many errors and for their suggestions for improved formatting that we subsequently incorporated. Hidefumi Tanaka assisted with some trials on the use of the ORACLE for Macintosh with HyperCard. We are grateful to him for permission to use some examples of output generated by him.

We are especially grateful to Phil McFadden who put forward the proposal for the Global Database originally at the IUGG meeting in Vancouver in 1987. Throughout the project he has been a constant source of advice on matters relating to PC systems and databases. He was responsible for setting up our own PC system at the beginning of the project and at the end he carefully read the manuscript and made many helpful suggestions for its improvement.

Appendix 1 – Table and View Descriptions

1.1. LIST OF TABLES AND VIEWS IN THE GPMDB

SQL⟩ SELECT TNAME, TABTYPE FROM TAB;

TNAME	TABTYPE
ALLCAGE	VIEW
ALTRESULT	TABLE
AUTHORS	TABLE
CROSSREF	TABLE
FIELDTESTS	TABLE
INFORMATION	TABLE
JOURNAL	TABLE
KEYS	TABLE
PMAGRESULT	TABLE
PRICAGE	VIEW
REFERENCE	TABLE
REMARKS	TABLE
ROCKUNIT	TABLE
RSLTALL	VIEW
RSLTPRI	VIEW
TIMESCALE	TABLE

1.2. DESCRIPTION OF TABLES IN THE GPMDB

SQL⟩ DESCRIBE ALTRESULT:

Name	Null?	Type
RESULTNO	NOT NULL	NUMBER(4)
APLAT	NOT NULL	NUMBER(3, 1)
APLONG	NOT NULL	NUMBER(4, 1)
KP		NUMBER(4, 1)
EP95		NUMBER(3, 1)

SQL⟩ DESCRIBE AUTHORS,

Name	Null?	Type
REFNO	NOT NULL	NUMBER(4)
AUTHORS	NOT NULL	CHAR(240)

SQL⟩ DESCRIBE CROSSREF;

Name	Null?	Type
RESULTNO	NOT NULL	NUMBER(4)
CATNO	NOT NULL	CHAR(10)

SQL⟩ DESCRIBE FIELDTESTS,

Name	Null?	Type
RESULTNO	NOT NULL	NUMBER(4)
TESTTYPE	NOT NULL	CHAR(2)
PARAMETERS	NOT NULL	CHAR(60)
SIGNIFICANCE		CHAR(3)

SQL) DESCRIBE INFORMATION

Name	Null?	Type
SYMBOL		CHAR(14)
EXPLANATION		CHAR(60)

SQL) DESCRIBE JOURNAL;

Name	Null?	Type
ABBREVIATION	NOT NULL	CHAR(8)
FULLNAME	NOT NULL	CHAR(60)

SQL) DESCRIBE KEYS;

Name	Null?	Type
KEY		CHAR(27)
FUNCTION		CHAR(47)

SQL) DESCRIBE PMAGRESULT.

Name	Null?	Type
ROCKUNITNO	NOT NULL	NUMBER(4)
RESULTNO	NOT NULL	NUMBER(4)
COMPONENT		CHAR(15)
LOMAGAGE	NOT NULL	NUMBER(4)
HIMAGAGE	NOT NULL	NUMBER(4)
TESTS	NOT NULL	CHAR(13)
TILT	NOT NULL	NUMBER(3)
SLAT	NOT NULL	NUMBER(3, 1)
SLONG	NOT NULL	NUMBER(4, 1)
B		NUMBER(4)
N		NUMBER(4)
DEC	NOT NULL	NUMBER(4, 1)
INC	NOT NULL	NUMBER(3, 1)
KD		NUMBER(4, 1)
ED95		NUMBER(3, 1)
PLAT	NOT NULL	NUMBER(3, 1)
PLONG	NOT NULL	NUMBER(4, 1)
PTYPE	NOT NULL	CHAR(1)
DP		NUMBER(3, 1)
DM		NUMBER(3, 1)
NOREVERSED	NOT NULL	CHAR(6)
ANTIPODAL		NUMBER(4, 1)
N_NORM		NUMBER(4)
D_NORM		NUMBER(4, 1)
I_NORM		NUMBER(3, 1)
K_NORM		NUMBER(4, 1)
ED_NORM		NUMBER(3, 1)
N_REV		NUMBER(4)
D_REV		NUMBER(4, 1)
I_REV		NUMBER(3, 1)
K_REV		NUMBER(4, 1)
ED_REV		NUMBER(3, 1)
DEMAGCODE	NOT NULL	NUMBER(1)

TREATMENT	NOT NULL	CHAR(5)
LABDETAILS		CHAR(60)
ROCKMAG		CHAR(60)
N_TILT		NUMBER(4)
D_UNCOR		NUMBER(4, 1)
I_UNCOR		NUMBER(3, 1)
K1		NUMBER(4, 1)
ED1		NUMBER(3, 1)
D_COR		NUMBER(4, 1)
I_COR		NUMBER(3, 1)
K2		NUMBER(4, 1)
ED2		NUMBER(3, 1)
COMMENTS		CHAR(60)

SQL ⟩ DESCRIBE REFERENCE;

Name	Null?	Type
REFNO	NOT NULL	NUMBER(4)
YEAR	NOT NULL	NUMBER(4)
JOURNAL	NOT NULL	CHAR(60)
VOLUME		NUMBER(4)
PAGES	NOT NULL	CHAR(11)
TITLE	NOT NULL	CHAR(200)

SQL ⟩ DESCRIBE REMARKS;

Name	Null?	Type
REFNO	NOT NULL	NUMBER(4)
REMARKS	NOT NULL	CHAR(13)

SQL ⟩ DESCRIBE ROCKUNIT;

Name	Null?	Type
REFNO	NOT NULL	NUMBER(4)
ROCKUNITNO	NOT NULL	NUMBER(4)
ROCKNAME	NOT NULL	CHAR(60)
PLACE	NOT NULL	CHAR(60)
CONTINENT		CHAR(15)
TERRANE		CHAR(30)
RLAT	NOT NULL	NUMBER(3, 1)
RLONG	NOT NULL	NUMBER(4, 1)
ROCKTYPE	NOT NULL	CHAR(40)
STRATA		CHAR(60)
STRATAGE	NOT NULL	CHAR(7)
LATSPREAD		CHAR(15)
LOWAGE	NOT NULL	NUMBER(4)
HIGHAGE	NOT NULL	NUMBER(4)
METHOD	NOT NULL	CHAR(30)
ISOTOPEDATA		CHAR(60)
STRUCTURE		CHAR(60)

SQL) DESCRIBE TIMESCALE;

Name	Null?	Type
PERIOD		CHAR(13)
SUBDIV		CHAR(9)
EPOCH		CHAR(14)
SYMBOL		CHAR(3)
BEGIN		CHAR(5)
END		CHAR(5)
NO		NUMBER(2)

1.3. LIST OF THE VIEWS IN THE GPMDB

SQL) DESCRIBE ALLCAGE;

Name	Null?	Type
ROCKNAME	NOT NULL	CHAR(60)
CONTINENT		CHAR(15)
PLACE	NOT NULL	CHAR(60)
LOMAGAGE	NOT NULL	NUMBER(4)
HIMAGAGE	NOT NULL	NUMBER(4)
TESTS	NOT NULL	CHAR(13)
DEMAGCODE	NOT NULL	NUMBER(1)
PLAT	NOT NULL	NUMBER(3, 1)
PLONG	NOT NULL	NUMBER(4, 1)
DP		NUMBER(3, 1)
DM		NUMBER(3, 1)

SQL) DESCRIBE PRICAGE;

Name	Null?	Type
ROCKNAME	NOT NULL	CHAR(60)
CONTINENT		CHAR(15)
PLACE	NOT NULL	CHAR(60)
LOMAGAGE	NOT NULL	NUMBER(4)
HIMAGAGE	NOT NULL	NUMBER(4)
TESTS	NOT NULL	CHAR(13)
DEMAGCODE	NOT NULL	NUMBER(1)
PLAT	NOT NULL	NUMBER(3, 1)
PLONG	NOT NULL	NUMBER(4, 1)
DP		NUMBER(3, 1)
DM		NUMBER(3, 1)

SQL) DESCRIBE RSLTALL;

Name	Null?	Type
AUTHORS	NOT NULL	CHAR(240)
YEAR	NOT NULL	NUMBER(4)
B		NUMBER(4)
N		NUMBER(4)
ROCKNAME	NOT NULL	CHAR(60)
CONTINENT		CHAR(15)
PLACE	NOT NULL	CHAR(60)

LOMAGAGE	NOT NULL	NUMBER(4)
HIMAGAGE	NOT NULL	NUMBER(4)
TESTS	NOT NULL	CHAR(13)
DEMAGCODE	NOT NULL	NUMBER(1)
PLAT	NOT NULL	NUMBER(3, 1)
PLONG	NOT NULL	NUMBER(4, 1)
DP		NUMBER(3, 1)
DM		NUMBER(3, 1)

SQL > DESCRIBE RSLTPRI;

Name	Null?	Type
AUTHORS	NOT NULL	CHAR(240)
YEAR	NOT NULL	NUMBER(4)
B		NUMBER(4)
N		NUMBER(4)
ROCKNAME	NOT NULL	CHAR(60)
CONTINENT		CHAR(15)
PLACE	NOT NULL	CHAR(60)
LOMAGAGE	NOT NULL	NUMBER(4)
HIMAGAGE	NOT NULL	NUMBER(4)
TESTS	NOT NULL	CHAR(13)
DEMAGCODE	NOT NULL	NUMBER(1)
PLAT	NOT NULL	NUMBER(3, 1)
PLONG	NOT NULL	NUMBER(4, 1)
DP		NUMBER(3, 1)
DM		NUMBER(3, 1)

1.4. DESCRIPTION OF TABLES IN THE COMPACT DATABASE – ABASE

Table REF:

COLUMN NAME	LENGTH	FORMAT
REFNO	5	INTEGER
AUTHORS	127	CHARACTER
YEAR	5	INTEGER
JOURNAL	60	CHARACTER
VOLUME	5	INTEGER
PAGES	11	CHARACTER
TITLE	200	CHARACTER

Table ROCKUNIT:

COLUMN NAME	LENGTH	FORMAT
REFNO	5	INTEGER
ROCKUNITNO	5	INTEGER
ROCKNAME	60	CHARACTER
PLACE	60	CHARACTER
CONTINENT	15	CHARACTER
ROCKTYPE	40	CHARACTER
STRATAGE	7	CHARACTER
LOWAGE	5	INTEGER
HIGHAGE	5	INTEGER
METHOD	30	CHARACTER

Table RESULT:

COLUMN NAME	LENGTH	FORMAT
REFNO	5	INTEGER
ROCKUNITNO	5	INTEGER
RESULTNO	5	INTEGER
COMPONENT	15	CHARACTER
LOMAGAGE	5	INTEGER
HIMAGAGE	5	INTEGER
TESTS	13	CHARACTER
TILT	4	INTEGER
SLAT	5	F5.1
SLONG	6	F6.1
B	5	INTEGER
N	5	INTEGER
DEC	6	F6.1
INC	5	F5.1
KD	6	F6.1
ED95	5	F5.1
PLAT	5	F5.1
PLONG	6	F6.1
DP	5	F5.1
DM	5	F5.1
NOREVERSED	6	CHARACTER
DEMAGCODE	2	INTEGER
TREATMENT	5	CHARACTER
COMMENTS	60	CHARACTER

Appendix 2 – View Code

1. SQL Statements Used to Create the GPMDB Views

1.1. VIEW ALLCAGE

```
CREATE VIEW ALLCAGE AS
SELECT ROCKNAME, CONTINENT, PLACE, LOMAGAGE, HIMAGAGE, TESTS,
       DEMAGCODE, PLAT, PLONG, DP, DM
FROM   ROCKUNIT RU, PMAGRESULT PMR
WHERE RU.ROCKUNITNO = PMR.ROCKUNITNO;
```

1.2. VIEW PRICAGE

```
CREATE VIEW PRICAGE AS
SELECT ROCKNAME, CONTINENT, PLACE, LOMAGAGE, HIMAGAGE, TESTS,
       DEMAGCODE, PLAT, PLONG, DP, DM
```

```
FROM     ROCKUNIT RU, PMAGRESULT PMR
WHERE RU.ROCKUNITNO  = PMR.ROCKUNITNO
AND     RU.LOWAGE     = PMR.LOMAGAGE
AND     RU.HIGHAGE    = PMR.HIMAGAGE;
```

1.3. VIEW RSLTALL

```
CREATE VIEW RSLTALL AS
SELECT AUTHORS, YEAR, B, N, ROCKNAME, CONTINENT, PLACE, LOMAGAGE,
        HIMAGAGE, TESTS, DEMAGCODE, PLAT, PLONG, DP, DM
FROM     AUTHORS A, REFERENCE R, ROCKUNIT RU, PMAGRESULT PMR
WHERE RU.ROCKUNITNO  = PMR.ROCKUNITNO
AND     RU.REFNO      = A.REFNO
AND     RU.REFNO      = R.REFNO
```

1.4. VIEW RSLTPRI

```
CREATE VIEW RSLTPRI AS
SELECT AUTHORS, YEAR, B, N, ROCKNAME, CONTINENT, PLACE, LOMAGAGE,
        HIMAGAGE, TESTS, DEMAGCODE, PLAT, PLONG, DP, DM
FROM     AUTHORS A, REFERENCE R, ROCKUNIT RU, PMAGRESULT PMR
WHERE RU.ROCKUNITNO  = PMR.ROCKUNITNO
AND     RU.REFNO      = A.REFNO
AND     RU.REFNO      = R.REFNO
AND     RU.LOWAGE     = PMR.LOMAGAGE
AND     RU.HIGHAGE    = PMR.HIMAGAGE;
```

Appendix 3 – Custombuilt .SQL Command File Code

```
SQL)    SET BUFFER SCRATCH;
SQL)    GET SCREEN.SQL;
    1    REM SCREEN CONFIGURATION COMMAND FILE
    2    SET NEWPAGE 1;
    3    SET PAGESIZE 25;
    4    SET LINESIZE 80;
    5    SET SPACE 1;
    6    SET ECHO OFF;
    7    SET TERMOUT ON;
    8    SET FEEDBACK ON;
    9    SET VERIFY OFF;
   10    SET PAUSE ON;
   11*   SET PAUSE "Press ENTER or RETURN to continue";

SQL)    GET PRINTER;
    1    REM PRINTER CONFIGURATION COMMAND FILE
    2    SET NEWPAGE 4;
    3    SET PAGESIZE 66;
    4    SET LINESIZE 80;
```

```
 5   SET PAUSE OFF;
 6   SET TERMOUT OFF:
 7   SET FEEDBACK OFF;
 8   SET ECHO OFF;
 9*  SET VERIFY OFF;
```

```
SQL)   GET SALLCAGE;
  1    REM COMMAND FILE THAT LISTS A SUMMARY OF PALEOMAGNETIC
  2    REM RESULTS FOR A USER SUPPLIED CONTINENT AND TIME WINDOW.
  3    REM RESULTS FOR BOTH PRIMARY AND SECONDARY MAGNETIZATION
  4    REM ARE SELECTED.
  5    SET NEWPAGE 1;
  6    SET PAGESIZE 25;
  7    SET LINESIZE 80;
  8    SET SPACE 1;
  9    SET TERMOUT ON;
 10    SET FEEDBACK ON;
 11    SET ECHO OFF;
 12    SET VERIFY OFF;
 13    SET PAUSE ON;
 14    SET PAUSE "Press ENTER or RETURN to continue";
 15    COLUMN ROCKNAME FORMAT A18 JUSTIFY C HEADING 'ROCK NAME' –
       WORD_WRAP;
 16    COLUMN CONTINENT NOPRINT NEW_VAL CONTVAR;
 17    COLUMN PLACE FORMAT A12 WORD_WRAP;
 18    COLUMN LOMAGAGE FORMAT 9990 HEADING
       'LOW|MAG|AGE';
 19    COLUMN HIMAGAGE FORMAT 9999 HEADING
       'HIGH|MAG|AGE';
 20    COLUMN TESTS FORMAT A6 WORD_WRAP;
 21    COLUMN DEMAGCODE FORMAT 0 HEADING 'D|M|C';
 22    COLUMN PLAT FORMAT 999.9 HEADING 'POLE|LAT'
 23    COLUMN PLONG FORMAT 999.9 HEADING 'POLE|LONG'
 24    COLUMN DP FORMAT 99.9 JUSTIFY C HEADING 'DP';
 25    COLUMN DM FORMAT 99.9 JUSTIFY C HEADING 'DM';
 26    BREAK ON CONTINENT;
 27    TTITLE LEFT 'CONTINENT: ' CONTVAR;
 28    SELECT ROCKNAME. CONTINENT. PLACE, LOMAGAGE,
 29    HIMAGAGE, TESTS, DEMAGCODE, PLAT, PLONG, DP, DM
 30    FROM ALLCAGE
 31    WHERE CONTINENT = '&CONTINENT'
 32    AND HIMAGAGE ) '&LOAGE'
 33*   AND LOMAGAGE < '&HIAGE';
```

```
SQL)   GET PALLCAGE;
  1    REM COMMAND FILE THAT LISTS A SUMMARY OF PALEOMAGNETIC
  2    REM RESULTS FOR A USER SUPPLIED CONTINENT AND TIME WINDOW.
  3    REM RESULTS FOR BOTH PRIMARY AND SECONDARY MAGNETIZATION
  4    REM ARE SELECTED.
  5    SET NEWPAGE 4;
  6    SET PAGESIZE 66;
  7    SET LINESIZE 80;
  8    SET SPACE 1;
  9    SET PAUSE OFF;
```

```
10   SET VERIFY OFF;
11   SET FEEDBACK OFF;
12   SET TERMOUT ON;
13   SET ECHO OFF;
14   COLUMN ROCKNAME FORMAT A18 JUSTIFY C HEADING 'ROCK NAME' –
     WORD_WRAP;
15   COLUMN CONTINENT NOPRINT NEW_VAL CONTVAR;
16   COLUMN PLACE FORMAT A12 WORD_WRAP;
17   COLUMN LOMAGAGE FORMAT 9990 HEADING 'LOW|MAG|AGE';
18   COLUMN HIMAGAGE FORMAT 9999 HEADING 'HIGH|MAG|AGE';
19   COLUMN TESTS FORMAT A6 WORD_WRAP;
20   COLUMN DEMAGCODE FORMAT 0 HEADING 'D|M|C';
21   COLUMN PLAT FORMAT 999.9 HEADING 'POLE|LAT';
22   COLUMN PLONG FORMAT 999.9 HEADING 'POLE|LONG';
23   COLUMN DP FORMAT 99.9 JUSTIFY C HEADING 'DP';
24   COLUMN DM FORMAT 99.9 JUSTIFY C HEADING 'DM';
25   BREAK ON CONTINENT;
26   TTITLE LEFT 'CONTINENT: ' CONTVAR;
27   BTITLE CENTER 'GONDWANA CONSULTANTS' SKIP 2;
28   SPOOL PALLCAGE;
29   SELECT ROCKNAME, CONTINENT, PLACE, LOMAGAGE,
30   HIMAGAGE, TESTS, DEMAGCODE, PLAT, PLONG, DP, DM
31   FROM ALLCAGE
32   WHERE CONTINENT  =  '&CONTINENT'
33   AND HIMAGAGE > '&LOAGE'
34   AND LOMAGAGE < '&HIAGE';
35   SPOOL OFF;
36*  HOST COPY PALLCAGE.LST LPT1;


SQL )  GET SPRICAGE;
  1    REM COMMAND FILE THAT LISTS A SUMMARY OF PALEOMAGNETIC
  2    REM RESULTS FOR A USER SUPPLIED CONTINENT AND TIME WINDOW.
  3    REM RESULTS FOR ONLY PRIMARY MAGNETIZATION ARE SELECTED.
  4    SET NEWPAGE 1;
  5    SET PAGESIZE 25;
  6    SET LINESIZE 80;
  7    SET SPACE 1;
  8    SET TERMOUT ON;
  9    SET FEEDBACK ON;
 10    SET ECHO OFF;
 11    SET VERIFY OFF;
 12    SET PAUSE ON;
 13    SET PAUSE "Press ENTER or RETURN to continue";
 14    COLUMN ROCKNAME FORMAT A18 JUSTIFY C HEADING 'ROCK NAME' –
       WORD_WRAP;
 15    COLUMN CONTINENT NOPRINT NEW_VAL CONTVAR;
 16    COLUMN PLACE FORMAT A12 WORD_WRAP;
 17    COLUMN LOMAGAGE FORMAT 9990 HEADING 'PRIM|LOW|AGE';
 18    COLUMN HIMAGAGE FORMAT 9999 HEADING 'PRIM|HIGH|AGE';
 19    COLUMN TESTS FORMAT A6 WORD WRAP;
 20    COLUMN DEMAGCODE FORMAT 0 HEADING 'D|M|C';
 21    COLUMN PLAT FORMAT 999.9 HEADING 'POLE|LAT';
 22    COLUMN PLONG FORMAT 999.9 HEADING 'POLE|LONG';
 23    COLUMN DP FORMAT 99.9 JUSTIFY C HEADING 'DP';
```

```
24     COLUMN DM FORMAT 99.9 JUSTIFY C HEADING 'DM';
25     BREAK ON CONTINENT;
26     TTITLE LEFT 'CONTINENT: ' CONTVAR;
27     SELECT ROCKNAME, CONTINENT, PLACE, LOMAGAGE,
28     HIMAGAGE, TESTS, DEMAGCODE, PLAT, PLONG, DP, DM
29     FROM PRICAGE
30     WHERE CONTINENT = '&CONTINENT'
31     AND HIMAGAGE > '&LOAGE'
32*    AND LOMAGAGE < '&HIAGE';


SQL )  GET PPRICAGE;
1      REM COMMAND FILE THAT LISTS A SUMMARY OF PALEOMAGNETIC
2      REM RESULTS FOR A USER SUPPLIED CONTINENT AND TIME WINDOW.
3      REM RESULTS FOR ONLY PRIMARY MAGNETIZATION ARE SELECTED.
4      SET NEWPAGE 4;
5      SET PAGESIZE 66;
6      SET LINESIZE 80;
7      SET SPACE 1;
8      SET PAUSE OFF;
9      SET VERIFY OFF;
10     SET FEEDBACK OFF;
11     SET TERMOUT ON;
12     SET ECHO OFF;
13     COLUMN ROCKNAME FORMAT A18 JUSTIFY C HEADING 'ROCK NAME' -
       WORD_WRAP;
14     COLUMN CONTINENT NOPRINT NEW_VAL CONTVAR;
15     COLUMN PLACE FORMAT A12 WORD_WRAP;
16     COLUMN LOMAGAGE FORMAT 9990 HEADING 'PRIM|LOW|AGE';
17     COLUMN HIMAGAGE FORMAT 9999 HEADING 'PRIM|HIGH|AGE';
18     COLUMN TESTS FORMAT A6 WORD_WRAP;
19     COLUMN DEMAGCODE FORMAT 0 HEADING 'D|M|C';
20     COLUMN PLAT FORMAT 999.9 HEADING 'POLE|LAT';
21     COLUMN PLONG FORMAT 999.9 HEADING 'POLE|LONG';
22     COLUMN DP FORMAT 99.9 JUSTIFY C HEADING 'DP';
23     COLUMN DM FORMAT 99.9 JUSTIFY C HEADING 'DM';
24     BREAK ON CONTINENT;
25     TTITLE LEFT 'CONTINENT: ' CONTVAR;
26     BTITLE CENTER 'GONDWANA CONSULTANTS' SKIP 2;
27     SPOOL PPRICAGE;
28     SELECT ROCKNAME, CONTINENT, PLACE, LOMAGAGE,
29     HIMAGAGE, TESTS, DEMAGCODE, PLAT, PLONG, DP, DM
30     FROM PRICAGE
31     WHERE CONTINENT = '&CONTINENT'
32     AND HIMAGAGE > '&LOAGE'
33     AND LOMAGAGE < '&HIAGE';
34     SPOOL OFF;
35*    HOST COPY PPRICAGE.LST LPT1;


SQL )  GET SRSLTALL;
1      REM LISTS THE SAME SUMMARY OF RESULTS AS SALLCAGE FOR A USER
2      REM SUPPLIED CONTINENT AND TIME WINDOW, WITH AN ADDITIONAL
3      REM LINE PRECEDING EACH RESULT LISTING THE AUTHORS, YEAR OF
4      REM THE REFERENCE AND B,N.
5      SET NEWPAGE 1;
```

```
 6   SET PAGESIZE 25;
 7   SET LINESIZE 80;
 8   SET SPACE 1;
 9   SET VERIFY OFF;
10   SET ECHO OFF;
11   SET TERMOUT ON;
12   SET FEEDBACK ON;
13   SET PAUSE ON;
14   SET PAUSE "Press ENTER or RETURN to continue";
15   COLUMN AUTHORS FORMAT A61 WORD_WRAP;
16   COLUMN YEAR FORMAT 9999 JUSTIFY R;
17   COLUMN B FORMAT 9999 JUSTIFY R;
18   COLUMN N FORMAT 9999 JUSTIFY R;
19   COLUMN ROCKNAME FORMAT A18 HEADING 'ROCK NAME' WORD_WRAP;
20   COLUMN CONTINENT NOPRINT NEW_VAL CONTVAR;
21   COLUMN PLACE FORMAT A12 WORD_WRAP;
22   COLUMN LOMAGAGE FORMAT 9990 HEADING 'LOAGE';
23   COLUMN HIMAGAGE FORMAT 9999 HEADING 'HIAGE';
24   COLUMN TESTS FORMAT A5 WORD_WRAP;
25   COLUMN DEMAGCODE FORMAT 0 HEADING 'DC';
26   COLUMN PLAT FORMAT 999.9 HEADING 'PLAT';
27   COLUMN PLONG FORMAT 999.9 HEADING 'PLONG';
28   COLUMN DP FORMAT 99.9 JUSTIFY C HEADING 'DP';
29   COLUMN DM FORMAT 99.9 JUSTIFY C HEADING 'DM';
30   BREAK ON CONTINENT;
31   TTITLE LEFT 'CONTINENT: ' CONTVAR;
32   SELECT AUTHORS, YEAR, B, N, ROCKNAME, CONTINENT, PLACE,
33   LOMAGAGE, HIMAGAGE, TESTS, DEMAGCODE, PLAT, PLONG, DP, DM
34   FROM RSLTALL
35   WHERE CONTINENT = '&CONTINENT'
36   AND HIMAGAGE > '&LOAGE'
37   AND LOMAGAGE < '& HIAGE';


SQL)  GET PRSLTALL;
 1   REM LISTS THE SAME SUMMARY OF RESULTS AS PALLCAGE FOR A USER
 2   REM SUPPLIED CONTINENT AND TIME WINDOW, WITH AN ADDITIONAL
 3   REM LINE PRECEDING EACH RESULT LISTING THE AUTHORS, YEAR OF
 4   REM THE REFERENCE AND B,N.
 5   SET NEWPAGE 4;
 6   SET PAGESIZE 66;
 7   SET LINESIZE 80;
 8   SET SPACE 1;
 9   SET VERIFY OFF;
10   SET FEEDBACK OFF;
11   SET TERMOUT ON;
12   SET ECHO OFF;
13   SET PAUSE OFF;
14   COLUMN AUTHORS FORMAT A61 WORD_WRAP;
15   COLUMN YEAR FORMAT 9999 JUSTIFY R;
16   COLUMN B FORMAT 9999 JUSTIFY R;
17   COLUMN N FORMAT 9999 JUSTIFY R;
18   COLUMN ROCKNAME FORMAT A18 HEADING 'ROCK NAME' WORD_WRAP;
19   COLUMN CONTINENT NOPRINT NEW_VAL CONTVAR;
20   COLUMN PLACE FORMAT A12 WORD_WRAP;
```

```
21   COLUMN LOMAGAGE FORMAT 9990 HEADING 'LOAGE';
22   COLUMN HIMAGAGE FORMAT 9999 HEADING 'HIAGE';
23   COLUMN TESTS FORMAT A6 WORD_WRAP;
24   COLUMN DEMAGCODE FORMAT 0 HEADING 'DC';
25   COLUMN PLAT FORMAT 999.9 HEADING 'PLAT';
26   COLUMN PLONG FORMAT 999.9 HEADING 'PLONG';
27   COLUMN DP FORMAT 99.9 JUSTIFY C HEADING 'DP';
28   COLUMN DM FORMAT 99.9 JUSTIFY C HEADING 'DM';
29   BREAK ON CONTINENT;
30   TTITLE LEFT 'CONTINENT: ' CONTVAR;
31   BTITLE CENTER 'GONDWANA CONSULTANTS' SKIP 2;
32   SPOOL PRSLTALL;
33   SELECT AUTHORS, YEAR, B, N, ROCKNAME, CONTINENT, PLACE,
34   LOMAGAGE, HIMAGAGE, TESTS, DEMAGCODE, PLAT, PLONG, DP, DM
35   FROM RSLTALL
36   WHERE CONTINENT = '&CONTINENT'
37   AND HIMAGAGE > '&LOAGE'
38   AND LOMAGAGE < '&HIAGE';
39   SPOOL OFF;
40*  HOST COPY PRSLTALL.LST LPT1;


SQL ) GET SRSLTPRI;
1    REM LISTS THE SAME SUMMARY OF RESULTS AS PPRICAGE FOR A USER
2    REM SUPPLIED CONTINENT AND TIME WINDOW, WITH AN ADDITIONAL
3    REM LINE PRECEDING EACH RESULT LISTING THE AUTHORS, YEAR OF
4    REM THE REFERENCE AND B,N.
5    SET NEWPAGE 1;
6    SET PAGESIZE 25;
7    SET LINESIZE 80;
8    SET SPACE 1;
9    SET VERIFY OFF;
10   SET ECHO OFF;
11   SET TERMOUT ON;
12   SET FEEDBACK ON;
13   SET PAUSE ON;
14   SET PAUSE "Press ENTER or RETURN to continue";
15   COLUMN AUTHORS FORMAT A61 WORD_WRAP;
16   COLUMN YEAR FORMAT 9999 JUSTIFY R;
17   COLUMN B FORMAT 9999 JUSTIFY R;
18   COLUMN N FORMAT 9999 JUSTIFY R;
19   COLUMN ROCKNAME FORMAT A18 HEADING 'ROCK NAME' WORD_WRAP;
20   COLUMN CONTINENT NOPRINT NEW_VAL CONTVAR;
21   COLUMN PLACE FORMAT A12 WORD_WRAP;
22   COLUMN LOMAGAGE FORMAT 9990 HEADING 'LOAGE';
23   COLUMN HIMAGAGE FORMAT 9999 HEADING 'HIAGE';
24   COLUMN TESTS FORMAT A6 WORD_WRAP;
25   COLUMN DEMAGCODE FORMAT 0 HEADING 'DC';
26   COLUMN PLAT FORMAT 999.9 HEADING 'PLAT';
27   COLUMN PLONG FORMAT 999.9 HEADING 'PLONG';
28   COLUMN DP FORMAT 99.9 JUSTIFY C HEADING 'DP';
29   COLUMN DM FORMAT 99.9 JUSTIFY C HEADING 'DM';
30   BREAK ON CONTINENT;
31   TTITLE LEFT 'CONTINENT: ' CONTVAR;
32   SELECT AUTHORS, YEAR, B, N, ROCKNAME, CONTINENT, PLACE,
```

```
33    LOMAGAGE, HIMAGAGE, TESTS, DEMAGCODE, PLAT, PLONG, DP, DM
34    FROM RSLTPRI
35    WHERE CONTINENT  =  '&CONTINENT'
36    AND HIMAGAGE  >  '&LOAGE'
37*   AND LOMAGAGE  <  '&HIAGE';


SQL)  GET PRSLTPRI;
 1    REM LISTS THE SAME SUMMARY OF RESULTS AS PPRICAGE FOR A USER
 2    REM SUPPLIED CONTINENT AND TIME WINDOW, WITH AN ADDITIONAL
 3    REM LINE PRECEDING EACH RESULT LISTING THE AUTHORS, YEAR OF
 4    REM THE REFERENCE AND B,N.
 5    SET NEWPAGE 4;
 6    SET PAGESIZE 66;
 7    SET LINESIZE 80;
 8    SET VERIFY OFF;
 9    SET PAUSE OFF;
10    SET ECHO OFF;
11    SET TERMOUT ON;
12    SET FEEDBACK OFF;
13    COLUMN AUTHORS FORMAT A61 WORD_WRAP;
14    COLUMN YEAR FORMAT 9999 JUSTIFY R;
15    COLUMN B FORMAT 9999 JUSTIFY R;
16    COLUMN N FORMAT 9999 JUSTIFY R;
17    COLUMN ROCKNAME FORMAT A18 HEADING 'ROCK NAME' WORD_WRAP;
18    COLUMN CONTINENT NOPRINT NEW_VAL CONTVAR;
19    COLUMN PLACE FORMAT A12 WORD_WRAP;
20    COLUMN LOMAGAGE FORMAT 9990 HEADING 'LOAGE';
21    COLUMN HIMAGAGE FORMAT 9999 HEADING 'HIAGE';
22    COLUMN TESTS FORMAT A6 WORD_WRAP;
23    COLUMN DEMAGCODE FORMAT 0 HEADING 'DC';
24    COLUMN PLAT FORMAT 999.9 HEADING 'PLAT';
25    COLUMN PLONG FORMAT 999.9 HEADING 'PLONG';
26    COLUMN DP FORMAT 99.9 JUSTIFY C HEADING 'DP';
27    COLUMN DM FORMAT 99.9 JUSTIFY C HEADING 'DM';
28    BREAK ON CONTINENT;
29    TTITLE LEFT 'CONTINENT: ' CONTVAR;
30    BTITLE CENTER 'GONDWANA CONSULTANTS' SKIP 2;
31    SPOOL PRSLTPRI;
32    SELECT AUTHORS, YEAR, B, N, ROCKNAME, CONTINENT, PLACE
33    LOMAGAGE, HIMAGAGE, TESTS, DEMAGCODE, PLAT, PLONG, DP, DM
34    FROM RSLTPRI
35    WHERE CONTINENT  =  '&CONTINENT'
36    AND HIMAGAGE  >  '&LOAGE'
37    AND LOMAGAGE  <  '&HIAGE';
38    SPOOL OFF;
39*   HOST COPY PRSLTPRI.LST LPT1;


SQL)  GET SREFLIST;
 1    REM THIS IS A COMMAND FILE THAT RETRIEVES REFERENCES FOR A
 2    REM USER SUPPLIED AUTHOR NAME AND OUTPUTS THEM TO THE SCREEN
 3    REM IN A FORMAT SIMILAR TO A STANDARD PUBLISHED REFERENCE.
 4    SET NEWPAGE 1;
 5    SET PAGESIZE 25;
 6    SET LINESIZE 80;
```

```
 7   SET SPACE 1;
 8   SET VERIFY OFF;
 9   SET ECHO OFF;
10   SET TERMOUT ON
11   SET FEEDBACK ON;
12   SET PAUSE ON;
13   SET PAUSE "Press ENTER or RETURN to continue";
14   COLUMN AUTHORS FORMAT A79 WORD_WRAP;
15   COLUMN YEAR FORMAT 9999 JUSTIFY R;
16   COLUMN JOURNAL FORMAT A32 WORD_WRAP;
17   COLUMN VOLUME FORMAT 9999 JUSTIFY R HEADING 'VOL';
18   COLUMN PAGES FORMAT A34;
19   COLUMN TITLE FORMAT A79 WORD_WRAP;
20   SELECT AUTHORS, YEAR, JOURNAL, VOLUME, PAGES, TITLE
21   FROM REFERENCE R, AUTHORS A
22   WHERE R.REFNO = A.REFNO
23   AND AUTHORS LIKE '%&AUTHORNAME%'
24*  ORDER BY YEAR, JOURNAL, VOLUME, PAGES;


SQL)   GET PREFLIST;
 1   REM THIS IS A COMMAND FILE THAT RETRIEVES REFERENCES FOR A
 2   REM USER SUPPLIED AUTHOR NAME AND OUTPUTS THEM TO THE
     PRINTER
 3   REM IN A FORMAT SIMILAR TO A STANDARD PUBLISHED REFERENCE.
 4   SET NEWPAGE 4;
 5   SET PAGESIZE 66;
 6   SET LINESIZE 80;
 7   SET SPACE 1;
 8   SET VERIFY OFF;
 9   SET ECHO OFF;
10   SET TERMOUT ON;
11   SET FEEDBACK OFF;
12   SET PAUSE OFF;
13   COLUMN AUTHORS FORMAT A79 WORD_WRAP;
14   COLUMN YEAR FORMAT 9999 JUSTIFY R;
15   COLUMN JOURNAL FORMAT A32 WORD_WRAP;
16   COLUMN VOLUME FORMAT 9999 JUSTIFY R HEADING 'VOL';
17   COLUMN PAGES FORMAT A34;
18   COLUMN TITLE FORMAT A79 WORD_WRAP;
19   BTITLE CENTER 'GONDWANA CONSULTANTS' SKIP 2;
20   SPOOL PREFLIST;
21   SELECT AUTHORS, YEAR, JOURNAL, VOLUME, PAGES, TITLE
22   FROM AUTHORS A, REFERENCE R
23   WHERE A.REFNO = R.REFNO
24   AND AUTHORS LIKE '%&AUTHORNAME%'
25   ORDER BY YEAR, JOURNAL, VOLUME, PAGES;
26   SPOOL OFF;
27*  HOST COPY PREFLIST.LST LPT1;


SQL)   GET SAUTHQRY;
 1   REM THIS IS A COMMAND FILE THAT RETRIEVES REFERENCES FOR A
 2   REM USER SUPPLIED AUTHOR NAME AND YEAR. IF THE EXACT YEAR IS
 3   REM UNKNOWN, ENTER THE DECADE BY OMITTING THE LAST DIGIT OF
 4   REM THE YEAR. OUTPUT TO THE SCREEN IS SIMILAR TO A STANDARD
```

```
 5   REM PUBLISHED REFERENCE.
 6   SET NEWPAGE 1;
 7   SET PAGESIZE 25;
 8   SET LINESIZE 80;
 9   SET SPACE 1;
10   SET VERIFY OFF;
11   SET ECHO OFF;
12   SET TERMOUT ON;
13   SET FEEDBACK ON;
14   SET PAUSE ON;
15   SET PAUSE "Press ENTER or RETURN to continue";
16   COLUMN AUTHORS FORMAT A79 WORD_WRAP;
17   COLUMN YEAR FORMAT 9999 JUSTIFY R;
18   COLUMN JOURNAL FORMAT A32 WORD_WRAP;
19   COLUMN VOLUME FORMAT 9999 JUSTIFY R HEADING 'VOL';
20   COLUMN PAGES FORMAT A34;
21   COLUMN TITLE FORMAT A79 WORD_WRAP;
22   SELECT AUTHORS, YEAR, JOURNAL, VOLUME, PAGES, TITLE
23   FROM REFERENCE R, AUTHORS A
24   WHERE R.REFNO = A.REFNO
25   AND AUTHORS LIKE '%&AUTHORNAME%'
26   AND YEAR LIKE '&YEAR%'
27*  ORDER BY YEAR, JOURNAL, VOLUME, PAGES;


SQL)  GET PAUTHQRY;
 1   REM THIS IS A COMMAND FILE THAT RETRIEVES REFERENCES FOR A
 2   REM USER SUPPLIED AUTHOR NAME AND YEAR. IF THE EXACT YEAR IS
 3   REM UNKNOWN ENTER THE DECADE BY OMITTING THE LAST DIGIT OF
 4   REM THE YEAR. OUTPUT TO THE PRINTER IS SIMILAR TO A
 5   REM STANDARD PUBLISHED REFERENCE.
 6   SET NEWPAGE 4;
 7   SET PAGESIZE 66;
 8   SET LINESIZE 80;
 9   SET SPACE 1;
10   SET VERIFY OFF;
11   SET ECHO OFF;
12   SET TERMOUT ON;
13   SET FEEDBACK OFF;
14   SET PAUSE OFF;
15   COLUMN AUTHORS FORMAT A79 WORD_WRAP;
16   COLUMN YEAR FORMAT 9999 JUSTIFY R;
17   COLUMN JOURNAL FORMAT A32 WORD_WRAP;
18   COLUMN VOLUME FORMAT 9999 JUSTIFY R HEADING 'VOL';
19   COLUMN PAGES FORMAT A34;
20   COLUMN TITLE FORMAT A79 WORD_WRAP;
21   BTITLE CENTER 'GONDWANA CONSULTANTS' SKIP 2;
22   SPOOL PAUTHQRY;
23   SELECT AUTHORS, YEAR, JOURNAL, VOLUME, PAGES, TITLE
24   FROM AUTHORS A, REFERENCE R
25   WHERE A.REFNO = R.REFNO
26   AND AUTHORS LIKE '%&AUTHORNAME%'
27   AND YEAR LIKE '&YEAR%'
28   ORDER BY YEAR, JOURNAL, VOLUME, PAGES
29   SPOOL OFF;
```

30*   HOST COPY PAUTHQRY.LST LPT1;

SQL) GET ACTIVITY;
    1   REM A COMMAND FILE THAT GIVES A SUMMARY PALEOMAGNETIC
    2   REM ACTIVITY FOR A USER SPECIFIED AUTHOR. A COUNT OF RESULTS,
    3   REM SITES SAMPLES AND THEIR MINIMUM AND MAXIMUM LATITUDE
    4   REM LONGITUDE RANGE IS GIVEN. A PRINT SPOOL FILE IS CREATED
    5   REM AND CAN BE PRINTED IF DESIRED.
    6   SET NEWPAGE 1;
    7   SET PAGESIZE 15;
    8   SET LINESIZE 75;
    9   SET SPACE 1;
   10   SET TERMOUT ON;
   11   SET FEEDBACK OFF;
   12   SET ECHO OFF;
   13   SET VERIFY OFF;
   14   SET PAUSE OFF;
   15   BREAK ON DUMMY;
   16   COLUMN CONTINENT FORMAT A14,
   17   COLUMN "DATA" FORMAT 999;
   18   COLUMN "SITES" FORMAT 9999;
   19   COLUMN "SAMPLES" FORMAT 99999;
   20   COLUMN "MIN LAT" FORMAT 99.9;
   21   COLUMN "MAX LAT" FORMAT 99.9;
   22   COLUMN "MIN LONG" FORMAT 999.9;
   23   COLUMN "MAX LONG" FORMAT 999.9;
   24   COMPUTE SUM OF "DATA" ON DUMMY;
   25   COMPUTE SUM OF "SITES" ON DUMMY;
   26   COMPUTE SUM OF "SAMPLES" ON DUMMY;
   27   COLUMN DUMMY NOPRINT;
   28   TTITLE 'SUMMARY OF PALEOMAGNETIC ACTIVITIES FOR NAMED
         AUTHOR' SKIP 2;
   29   SPOOL ACTIVITY;
   30   SELECT CONTINENT,
   31   COUNT(RESULTNO) "DATA",
   32   SUM(B) "SITES",
   33   SUM(N) "SAMPLES",
   34   MIN(RLAT) "MIN LAT",
   35   MAX(RLAT) "MAX LAT",
   36   MIN(RLONG) "MIN LONG",
   37   MAX(RLONG) "MAX LONG",
   38   SUM(O) DUMMY
   39   FROM AUTHORS A, ROCKUNIT RU, PMAGRESULT PMR
   40   WHERE A.REFNO = RU.REFNO
   41   AND RU.ROCKUNITNO = PMR.ROCKUNITNO
   42   AND AUTHORS LIKE '%&AUTHOR%'
   43   GROUP BY CONTINENT;
   44*   SPOOL OFF;

SQL) GET STIME;
    1   REM COMMAND FILE THAT LISTS THE GEOLOGICAL TIME SCALE 1989
    2   REM ON THE SCREEN.
    3   SET NEWPAGE 1;
    4   SET PAGESIZE 25;

```
 5   SET LINESIZE 75;
 6   SET SPACE 5;
 7   SET TERMOUT ON;
 8   SET FEEDBACK OFF;
 9   SET ECHO OFF;
10   SET VERIFY OFF;
11   SET PAUSE ON;
12   SET PAUSE "Press ENTER or RETURN to continue";
13   COLUMN PERIOD FORMAT A13 JUSTIFY C HEADING 'PERIOD';
14   COLUMN SUBDIV FORMAT A9 HEADING ' SUB|DIVISION';
15   COLUMN EPOCH FORMAT A14 JUSTIFY C HEADING 'EPOCH';
16   COLUMN SYMBOL FORMAT A3 HEADING 'SYM|BOL';
17   COLUMN BEGIN FORMAT A5 JUSTIFY C HEADING 'TOP';
18   COLUMN END FORMAT A5 HEADING 'BASE';
19   TTITLE CENTER 'GEOLOGICAL TIMESCALE 1989';
20   SELECT PERIOD, SUBDIV, EPOCH, SYMBOL, BEGIN, END
21   FROM TIMESCALE
22*  ORDER BY NO;
```

```
SQL)  GET PTIME;
 1   REM COMMAND FILE THAT DIRECTS THE GEOLOGICAL TIME SCALE 1989
 2   REM TO THE PRINTER.
 3   SET NEWPAGE 1;
 4   SET PAGESIZE 66;
 5   SET LINESIZE 75;
 6   SET SPACE 5;
 7   SET TERMOUT OFF;
 8   SET FEEDBACK OFF;
 9   SET ECHO OFF;
10   SET VERIFY OFF;
11   SET PAUSE OFF;
12   COLUMN PERIOD FORMAT A13 JUSTIFY C HEADING 'PERIOD';
13   COLUMN SUBDIV FORMAT A9 HEADING ' SUB|DIVISION';
14   COLUMN EPOCH FORMAT A14 JUSTIFY C HEADING 'EPOCH';
15   COLUMN SYMBOL FORMAT A3 HEADING 'SYM|BOL';
16   COLUMN BEGIN FORMAT A5 JUSTIFY C HEADING 'TOP';
17   COLUMN END FORMAT A5 HEADING 'BASE';
18   TTITLE CENTER 'GEOLOGICAL TIMESCALE 1989' SKIP 2;
19   BTITLE CENTER 'GONDWANA CONSULTANTS';
20   SPOOL PTIME;
21   SELECT PERIOD, SUBDIV, EPOCH, SYMBOL, BEGIN, END
22   FROM  TIMESCALE
23   ORDER BY NO;
24   SPOOL OFF;
25*  HOST COPY PTIME.LST LPT1;
```

```
SQL)  GET SKEYS;
 1   REM THIS A COMMAND FILE THAT LISTS THE FUNCTION KEYS FOR USE
 2   REM WITH THE CUSTOMISED FORM PALEOMAG
 3   SET NEWPAGE 1;
 4   SET PAGESIZE 25;
 5   SET LINESIZE 79;
 6   SET SPACE 2;
 7   SET TERMOUT ON;
```

```
 8   SET VERIFY OFF;
 9   SET ECHO OFF;
10   SET FEEDBACK OFF;
11   SET PAUSE ON;
12   SET PAUSE "Press ENTER or RETURN to continue";
13   COLUMN KEY HEADING 'OPERATOR''S KEY';
14   TTITLE CENTER 'LIST OF KEYS FOR PALEOMAG FORM' SKIP 2;
15*  SELECT * FROM KEYS;
```

```
SQL)  GET PKEYS;
 1   REM THIS IS A COMMAND FILE THAT LISTS THE FUNCTION KEYS FOR
 2   REM USE WITH THE CUSTOMISED FORM PALEOMAG. OUTPUT IS
 3   REM AUTOMATICALLY DIRECTED TO THE PRINTER IF SWITCHED ON.
 4   SET NEWPAGE 1;
 5   SET PAGESIZE 66;
 6   SET LINESIZE 79;
 7   SET SPACE 2;
 8   SET FEEDBACK OFF;
 9   SET ECHO OFF
10   SET TERMOUT OFF;
11   SET VERIFY OFF;
12   SET PAUSE OFF;
13   SET SPACE 2;
14   COLUMN KEY HEADING 'OPERATOR''S KEY';
15   SPOOL PKEYS;
16   TTITLE CENTER 'LIST OF KEYS FOR PALEOMAG FORM' SKIP 2;
17   BTITLE CENTER 'GONDWANA CONSULTANTS';
18   SELECT * FROM KEYS;
19   SPOOL OFF;
20*  HOST COPY PKEYS.LST LPT1;
```

```
SQL)  GET SJOURNAL;
 1   REM THIS IS A COMMAND FILE THAT LISTS JOURNAL ENTRY STYLES
 2   REM USED IN THE REFERENCE TABLE OF THE GPMDB.
 3   SET NEWPAGE 1;
 4   SET PAGESIZE 25;
 5   SET LINESIZE 79;
 6   SET SPACE 2;
 7   SET TERMOUT ON;
 8   SET VERIFY OFF;
 9   SET ECHO OFF;
10   SET FEEDBACK OFF;
11   SET PAUSE ON;
12   SET PAUSE "Press ENTER or RETURN to continue";
13   COLUMN FULLNAME FORMAT A50;
14   TTITLE CENTER 'JOURNAL STYLES' SKIP 2;
15   SELECT * FROM JOURNAL
16*  ORDER BY ABBREVIATION;
```

```
SQL)  GET PJOURNAL;
 1   REM THIS IS A COMMAND FILE THAT LISTS THE JOURNAL ENTRY
 2   REM STYLES USED IN THE REFERENCE TABLE OF GPMDB. OUTPUT IS
 3   REM AUTOMATICALLY DIRECTED TO THE PRINTER IF IT IS SWITCHED
```

```
 4    REM ON.
 5    SET NEWPAGE 1;
 6    SET PAGESIZE 66;
 7    SET LINESIZE 79;
 8    SET SPACE 2;
 9    SET FEEDBACK OFF;
10    SET ECHO OFF;
11    SET TERMOUT OFF;
12    SET VERIFY OFF;
13    SET PAUSE OFF;
14    COLUMN FULLNAME FORMAT A50;
15    SPOOL PJOURNAL;
16    TTITLE CENTER 'JOURNAL STYLES' SKIP 2;
17    BTITLE CENTER 'GONDWANA CONSULTANTS';
18    SELECT * FROM JOURNAL
19    ORDER BY ABBREVIATION;
20    SPOOL OFF;
21*   HOST COPY PJOURNAL.LST LPT1;


SQL>  GET SINFO;
 1    REM THIS IS A COMMAND FILE THAT LISTS EXPLANATIONS OF THE
 2    REM SYMBOLS USED IN THE GPMDB.
 3    SET NEWPAGE 1;
 4    SET PAGESIZE 25;
 5    SET LINESIZE 79;
 6    SET SPACE 2;
 7    SET TERMOUT ON;
 8    SET VERIFY OFF;
 9    SET ECHO OFF;
10    SET FEEDBACK OFF;
11    SET PAUSE ON;
12    SET PAUSE "Press ENTER or RETURN to continue";
13    COLUMN SYMBOL HEADING 'COLUMN/SYMBOL';
14    TTITLE CENTER 'INFORMATION' SKIP 2;
15    SELECT * FROM INFORMATION
16*   ORDER BY SYMBOL;


SQL>  GET PINFO;
 1    REM THIS A COMMAND FILE THAT LISTS EXPLANATIONS OF THE
 2    REM SYMBOLS USED IN THE GPMDB. OUTPUT IS AUTOMATICALLY
 3    REM DIRECTED TO THE PRINTER IF IT IS SWITCHED ON.
 4    SET NEWPAGE 1;
 5    SET PAGESIZE 66;
 6    SET LINESIZE 79;
 7    SET SPACE 2;
 8    SET FEEDBACK OFF;
 9    SET ECHO OFF;
10    SET TERMOUT OFF;
11    SET VERIFY OFF;
12    SET PAUSE OFF;
13    SPOOL PINFO;
14    COLUMN SYMBOL HEADING 'COLUMN/SYMBOL';
15    TTITLE CENTER 'INFORMATION' SKIP 2;
16    BTITLE CENTER 'GONDWANA CONSULTANTS' SKIP 6;
```

```
17   SELECT * FROM INFORMATION
18   ORDER BY SYMBOL;
19   SPOOL OFF;
20*  HOST COPY PINFO.LST LPT1;


SQL )   GET SALLVIEW;
  1   REM THIS IS A COMMAND FILE THAT LISTS YOUR VIEW NAMES AND
  2   REM THE SQL CODE USED TO CREATE THEM.
  3   SET NEWPAGE 1;
  4   SET PAGESIZE 25;
  5   SET LINESIZE 80;
  6   SET PAUSE ON;
  7   SET PAUSE "Press RETURN to continue";
  8   COLUMN VIEWNAME FORMAT A12;
  9   COLUMN VIEWTEXT FORMAT A66 WORD_WRAP;
 10   TTITLE CENTER 'LIST OF VIEWS OWNED BY THE USER' SKIP 2;
 11   SET LONG 900;
 12*  SELECT * FROM VIEWS;


SQL )   GET PALLVIEW;
  1   REM THIS IS A COMMAND FILE THAT LISTS YOUR VIEW NAMES AND
  2   REM SQL CODE USED TO CREATE THEM. OUTPUT IS AUTOMATICALLY
  3   REM DIRECTED TO THE PRINTER IF IT IS SWITCHED ON.
  4   SET NEWPAGE 4;
  5   SET PAGESIZE 66;
  6   SET LINESIZE 80;
  7   SET PAUSE OFF;
  8   SET FEEDBACK OFF;
  9   SET ECHO OFF;
 10   COLUMN VIEWNAME FORMAT A12;
 11   COLUMN VIEWTEXT FORMAT A66 WORD_WRAP;
 12   TTITLE CENTER 'LIST OF VIEWS OWNED BY THE USER' SKIP 2;
 13   BTITLE CENTER 'GONDWANA CONSULTANTS';
 14   SET LONG 900;
 15   SPOOL PALLVIEW;
 16   SELECT * FROM VIEWS;
 17   SPOOL OFF;
 18*  HOST COPY PALLVIEW.LST LPT1;


SQL )   GET SONEVIEW;
  1   REM THIS IS A COMMAND FILE THAT LISTS THE CODE THAT CREATED
  2   REM YOUR NAMED VIEW.
  3   SET NEWPAGE 1;
  4   SET PAGESIZE 25;
  5   SET LINESIZE 80;
  6   SET FEEDBACK OFF;
  7   SET ECHO OFF;
  8   SET PAUSE ON;
  9   SET PAUSE "Press RETURN to continue";
 10   COLUMN VIEWNAME FORMAT A12;
 11   COLUMN VIEWTEXT FORMAT A66 WORD WRAP;
 12   TTITLE CENTER 'CODE FOR A USER NAMED VIEW' SKIP 2;
 13   SET LONG 900;
```

```
14    SELECT * FROM VIEWS
15*   WHERE VIEWNAME  =  '&viewname';
```

```
SQL )    GET PONEVIEW;
    1    REM THIS IS A COMMAND FILE THAT LISTS THE CODE THAT CREATED
    2    REM YOUR NAMED VIEW. OUTPUT IS AUTOMATICALLY DIRECTED TO
    3    REM TO THE PRINTER IF IT IS SWITCHED ON.
    4    SET NEWPAGE 4;
    5    SET PAGESIZE 66;
    6    SET LINESIZE 80;
    7    SET FEEDBACK OFF;
    8    SET ECHO OFF;
    9    SET VERIFY OFF;
   10    SET PAUSE OFF;
   11    COLUMN VIEWNAME FORMAT A12;
   12    COLUMN VIEWTEXT FORMAT A66 WORD_WRAP;
   13    TTITLE CENTER 'CODE FOR A USER NAMED VIEW' SKIP 2;
   14    BTITLE CENTER 'GONDWANA CONSULTANTS';
   15    SET LONG 900;
   16    SPOOL PONEVIEW;
   17    SELECT * FROM VIEWS
   18    WHERE VIEWNAME  =  '&viewname';
   19    SPOOL OFF;
   20*   HOST COPY PONEVIEW.LST LPT1;
```

```
SQL )    GET SINDEXES;
    1    REM THIS IS A COMMAND FILE THAT LISTS YOUR INDEX NAMES, THE
    2    REM TABLES AND COLUMNS INDEXED AND THE TYPE OF INDEX.
    3    SET NEWPAGE 1;
    4    SET PAGESIZE 25;
    5    SET LINESIZE 80;
    6    SET ECHO OFF;
    7    SET TERMOUT ON;
    8    SET FEEDBACK ON;
    9    SET VERIFY OFF;
   10    SET PAUSE ON;
   11    SET PAUSE "Press ENTER or RETURN to continue";
   12    COLUMN INAME FORMAT A10 HEADING 'INDEX|NAME';
   13    COLUMN ICREATOR FORMAT A10 HEADING 'INDEX|CREATOR';
```

```
14   COLUMN TNAME FORMAT A11 HEADING 'TABLE|NAME';
15   COLUMN CREATOR FORMAT A10 HEADING 'TABLE|CREATOR';
16   COLUMN COLNAMES FORMAT A11 HEADING 'COLUMNS|INDEXED';
17   COLUMN INDEXTYPE FORMAT A10 HEADING'INDEX|TYPE';
18   COLUMN COMPRESSION FORMAT A10 HEADING –
     'COMPRESSED|NOCOMPRESS';
19   TTITLE 'YOUR INDEX NAMES, TABLES AND COLUMNS INDEXED '–
     'AND TYPE OF INDEX' SKIP 2;
20   SELECT INAME, ICREATOR, TNAME, CREATOR,
     COLNAMES, INDEXTYPE, COMPRESSION
21   FROM INDEXES
22*  ORDER BY TNAME, COLNAMES;
```

```
SQL )  GET PINDEXES;
  1    THIS IS A COMMAND FILE THAT LISTS YOUR INDEX NAMES, THE
  2    REM TABLES AND COLUMNS INDEXED AND THE TYPE OF INDEX. OUTPUT
  3    REM IS AUTOMATICALLY DIRECTED TO THE PRINTER IF IT IS
  4    REM SWITCHED ON.
  5    SET NEWPAGE 4;
  6    SET PAGESIZE 66;
  7    SET LINESIZE 80;
  8    SET ECHO OFF;
  9    SET TERMOUT ON;
 10    SET FEEDBACK OFF;
 11    SET VERIFY OFF;
 12    SET PAUSE OFF;
 13    COLUMN INAME FORMAT A10 HEADING 'INDEX|NAME';
 14    COLUMN ICREATOR FORMAT A10 HEADING 'INDEX|CREATOR';
 15    COLUMN TNAME FORMAT A11 HEADING 'TABLE|NAME';
 16    COLUMN CREATOR FORMAT A10 HEADING 'TABLE|CREATOR';
 17    COLUMN COLNAMES FORMAT A11 HEADING 'COLUMNS|INDEXED';
 18    COLUMN INDEXTYPE FORMAT A10 HEADING 'INDEX|TYPE';
 19    COLUMN COMPRESSION FORMAT A10 HEADING –
       'COMPRESSED|NOCOMPRESS';
 20    SPOOL PINDEXES;
 21    TTITLE 'INDEX NAMES, TABLES AND COLUMNS INDEXED AND' –
       'THE TYPE OF INDEX' SKIP 2;
 22    BTITLE CENTER 'GONDWANA CONSULTANTS';
 23    SELECT INAME, ICREATOR, TNAME, CREATOR,
       COLNAMES, INDEXTYPE, COMPRESSION
 24    FROM INDEXES;
 25    ORDER BY TNAME, COLNAMES;
 26    SPOOL OFF;
 27*   HOST COPY PINDEXES.LST LPT1;
```

Appendix 4 – SQL*Menu Application Documentation PMAG

A.1   Application Information
             Application   PMAG
             Creation date   19–NOV–90
             Creator   SYSTEM
             Version   1.1

A.2   Menu overview.
             BGM
             PMAG
               SPlus
               PPlus
               ORATU
               FORMS

A.3   BGM

BACKGROUND MENU

For the Global Paleomagnetic Database

1   List the Menu Name and corresponding Titles
2   Return to Application Menu

Gondwana Consultants

To list the menu names and corresponding titles

A.3.1   Option number   1
             Work class range   0–15
             Command type   SQL*Plus
             Command line
             SQLPLUS -s &UN/&PW @C:\ORACLE5\DMU\LISTMENU PMAG

A.3.2   Option number   2
             Work class range   0–15
             Command type   Macro command
             Command line   APLMENU;

A.4   FORMS

SQL*Forms Options

---

1   Use the Form PALEOMAG to Query the Whole Database
2   Use the Form KEYS to check Function Keys for PALEOMAG Form
3   Use the Form TIME to view the 1989 Geologic Time Scale
4   Use the Form JOURNAL to see standard formats
5   Use the Form INFO to check symbols
6   Previous Menu
7   Exit

Press HELP (F2) for further information

To RUN various prepackaged FORMS

A.4.1   Option number   1
            Work class range   0–15
            Command type   Runforms (IAP)
            Command line   RUNFORM PALEOMAG

A.4.2   Option number   2
            Work class range   0–15
            Command type   Runforms (IAP)
            Command line   RUNFORM KEYS

A.4.3   Option number   3
            Work class range   0–15
            Command type   Runforms (IAP)
            Command line   RUNFORM TIME

A.4.4   Option number   4
            Work class range   0–15
            Command type   Runforms (IAP)
            Command line   RUNFORM JOURNAL

A.4.5   Option number   5
            Work class range   0–15
            Command type   Runforms (IAP)
            Command line   RUNFORM INFO

A.4.6   Option number   6
            Work class range   0–15
            Command type   Macro command
            Command line   PRVMENU;

A.4.7   Option number   7
            Work class range   0–15

Command type	Macro command
Command lIne	EXIT;

A.5   ORATU

Oracle Tools and Utilities

---

1 SQL*Plus
2 Run a DOS Command
3 Run a FORM
4 SQL*Forms (Designer only)
5 Export
6 Import
7 Previous Menu
8 Exit

Press HELP (F2) for further information

To run Oracle Tools and Utilities directly from SQL*Menu

A.5.1   Option number   1
           Work class range   0–15
           Command type   SQL*Plus
           Command line   SQLPLUS &UN/&PW

A.5.2   Option number   2
           Work class range   0–15
           Command type   Macro command
           Command line   OSCMD;

A.5.3   Option number   3
           Work class range   0–15
           Command type   Runforms (IAP)
           Command line   RUNFORM &FN

A.5.4   Option number   4
           Work class range   0–15
           Command type   Macro command
           Command line   OSCMD SQLFORMS &UN/&PW;

A.5.5   Option number   5
           Work class range   0–15
           Command type   Macro command
           Command line   OSCMD EXP &UN/&PW;

A.5.6   Option number   6
           Work class range   0–15
           Command type   Macro command

	Command line	OSCMD IMP &UN/&PW;
A.5.7	Option number	7
	Work class range	0–15
	Command type	Macro command
	Command line	PRVMENU;
A.5.8	Option number	8
	Work class range	0–15
	Command type	Macro command
	Command line	EXIT;

A.6   PMAG

GLOBAL PALEOMAGNETIC DATABASE

MAIN MENU

1 SQL*Forms Options
2 SQL*Plus Screen Options
3 SQL*Plus Printer Options
4 Oracle Tools and Utilities
5 Previous Menu
6 Exit

Gondwana Consultants

To run various command files and utilities

A.6.1	Option number	1
	Work class range	0–15
	Command type	Menu Selection
	Command line	Forms
A.6.2	Option number	2
	Work class range	0–15
	Command type	Menu Selection
	Command line	SPlus
A.6.3	Option number	3
	Work class range	0–15
	Command type	Menu Selection
	Command line	PPlus
A.6.4	Option number	4
	Work class range	0–15
	Command type	Menu Selection
	Command line	OraTU

A.6.5   Option number   5
                   Work class range   0–15
                   Command type   Macro command
                   Command line   PRVMENU;

A.6.6   Option number   6
                   Work class range   0–15
                   Command type   Macro command
                   Command line   EXIT;

A.7   PPLUS

SQL*Plus Printer Options

_____

 1 Basic Pole Data by Continent and Age – All magnetizations
 2 Basic Pole Data by Continent and Age – Primary magnetizations
 3 Extended Pole Data by Continent and Age – All magnetizations
 4 Extended Pole Data by Continent and Age – Primary magnetizations
 5 Reference List for a given Author
 6 Listing of the 1989 Geologic Time Scale
 7 Listing of Function Keys for use with PALEOMAG Form
 8 Listing of Explanation of Symbols used in the Global Database
 9 Listing of Journal styles used in the Global Database
10 Previous Menu
11 Exit

Press HELP (F2) for further information

To run various SQL Command files

A.7.1   Option number   1
                   Work class range   0–15
                   Command type   SQL*Plus
                   Command line   SQLPLUS &UN/&PW @PALLCAGE

A.7.2   Option number   2
                   Work class range   0–15
                   Command type   SQL*Plus
                   Command line   SQLPLUS &UN/&PW @PPRICAGE

A.7.3   Option number   3
                   Work class range   0–15
                   Command type   SQL*Plus
                   Command line   SQLPLUS &UN/&PW @PRSLTALL

A.7.4   Option number   4
                   Work class range   0–15
                   Command type   SQL*Plus

	Command line	SQLPLUS &UN/&PW @PRSLTPRI
A.7.5	Option number	5
	Work class range	0–15
	Command type	SQL*Plus
	Command line	SQLPLUS &UN/&PW @PREFLIST
A.7.6	Option number	6
	Work class range	0–15
	Command type	SQL*Plus
	Command line	SQLPLUS &UN/&PW @PTIME
A.7.7	Option number	7
	Work class range	0–15
	Command type	SQL*Plus
	Command line	SQLPLUS &UN/&PW @PKEYS
A.7.8	Option number	8
	Work class range	0–15
	Command type	SQL*Plus
	Command line	SQLPLUS &UN/&PW @PINFO
A.7.9	Option number	9
	Work class range	0–15
	Command type	SQL*Plus
	Command line	SQLPLUS &UN/&PW @PJOURNAL
A.7.10	Option number	10
	Work class range	0–15
	Command type	Macro command
	Command line	PRVMENU;
A.7.11	Option number	11
	Work class range	0–15
	Command type	Macro command
	Command line	EXIT;

A.8   SPLUS

SQL*Plus Screen Options

---

1   Basic Pole Data by Continent and Age – All magnetizations
2   Basic Pole Data by Continent and Age – Primary magnetizations
3   Extended Pole Data by Continent and Age – All magnetizations
4   Extended Pole Data by Continent and Age – Primary magnetizations
5   Reference List for a given Author
6   Search for Reference – Author spelling and Year uncertain
7   Listing of 1989 Geologic Time Scale
8   Listing of Function Keys used in the PALEOMAG Form
9   Listing of Explanation of Symbols used in the Global Database
10 Previous Menu
11 Exit

Press HELP (F2) for further information

To run various SQL Command Files directly to the Screen

A.8.1   Option number   1
          Work class range   0–15
          Command type   SQL*Plus
          Command line   SQLPLUS &UN/&PW @SALLCAGE

A.8.2   Option number   2
          Work class range   0–15
          Command type   SQL*Plus
          Command line   SQLPLUS &UN/&PW @SPRICAGE

A.8.3   Option number   3
          Work class range   0–15
          Command type   SQL*Plus
          Command line   SQLPLUS &UN/&PW @SRSLTALL

A.8.4   Option number   4
          Work class range   0–15
          Command type   SQL*Plus
          Command line   SQLPLUS &UN/&PW @SRSLTPRI

A.8.5   Option number   5
          Work class range   0–15
          Command type   SQL*Plus
          Command line   SQLPLUS &UN/&PW @SREFLIST

A.8.6   Option number   6
          Work class range   0–15
          Command type   SQL*Plus

	Command line	SQLPLUS &UN/&PW @SAUTHQRY

A.8.7	Option number	7
	Work class range	0–15
	Command type	SQL*Plus
	Command line	SQLPLUS &UN/&PW @STIME

A.8.8	Option number	8
	Work class range	0–15
	Command type	SQL*Plus
	Command line	SQLPLUS &UN/&PW @SKEYS

A.8.9	Option number	9
	Work class range	0–15
	Command type	SQL*Plus
	Command line	SQLPLUS &UN/&PW @SINFO

A.8.10	Option number	10
	Work class range	0–15
	Command type	Macro command
	Command line	PRVMENU;

A.8.11	Option number	11
	Work class range	0–15
	Command type	Macro command
	Command line	EXIT;

A.9	Work Class	5
	General Application User	

Username	BG Menu	OS Command	Debug
WORKSHOP	N	Y	Y

A.10	Work Class	15
	DBA for this application only	

Username	BG Menu	OS Command	Debug
DATABASE	Y	Y	Y

A.11	Parameters

Parameter:	Prompt:
FN	FORM NAME

Parameter: FN
Attributes:

Echo:	Y
Must fill:	N
Response:	Y
Uppercase:	Y